PHENETICS: EVOLUTION, POPULATION, TRAIT

Phenetics:
Evolution, Population, Trait

A. V. Yablokov

Translated by MARIE JAROSZEWSKA HALL

COLUMBIA UNIVERSITY PRESS

New York

1986

Columbia University Press
New York   Guildford, Surrey
Fenetika: copyright © 1980 Nauka Publishers: Moscow
English edition: copyright © 1986 Columbia University Press

Printed in the United States of America

Library of Congress Cataloging-in-Publication Data

Yablokov, A. V. (Alekseĭ Vladimirovich)
    Phenetics—evolution, population, trait.

    Translated from the Russian.
    Bibliography: p.
    Includes index.
    1. Phenetics.   I. Title.
QH408.5.I24   1986   575   85-31420
ISBN 978-0-231-05990-9

This book is Smyth-sewn.
Book design by J. S. Roberts

Contents

Preface

To my American readers:

An increasing number of biologists from various countries are reaching the conclusion that the study of natural populations is of primary importance for understanding the patterns of change in the earth's biosphere. Understanding these patterns in the living world is basic to a better organization of measures to conserve species diversity in nature—one of the important problems confronting mankind—and to exploit living resources rationally.

Unfortunately, many of these patterns are still far from clear to us, and the reason for this is the practical impossibility of genetic study of any significant number of species under natural conditions. The phenetic approach, briefly described in this book, shows promise for opening unprecedented, broad possibilities for genetic investigation of natural populations by taking account of phenes, traits-markers of genetic composition.

In this book I cite much data from Russian scientific literature probably not especially well known in the United States. I have drawn as much data from Western literature for supporting many positions in phenetics. Such a combination of knowledge is important and indispensable: not one of the divisions of modern biology, in my opinion, can develop successfuly on a national basis only. For this reason I am indebted to my American colleagues who took the initiative to translate this book into English and to Columbia University Press for publishing it. As a result of this publication, the circle of investigators deliberately using phenetic approaches and methods will probably expand, and this will aid the further development of all of population biology.

In this, the third edition (the first was published in Moscow in 1980, the second in Tokyo in 1982), certain figures have been replaced and new ones added, certain (the majority, I hope) unfortunate imprecisions have been eliminated, and some aspects of phenetics have been presented in greater detail.

Moscow, April 1986   Aleksey Yablokov

Introduction

This book is the story of one of the new directions in the study of natural populations of animals and plants. In our time the study of a population under natural conditions is of interest to botanists and zoologists, ecologists and geneticists, morphologists and physiologists. The study of a population is important for agriculture, fisheries and game management, industry, medicine, and nature conservation, as well as for architects, sociologists, and many other specialists who until recently remained aloof from biological problems in general.

Will mankind of the future scurry about under transparent domes of megalopolises surrounded by industrial wastelands, or will it harmoniously coexist with the varied, rich, and living natural world that gives people all that is necessary for life and is continuously enriched under the control and in the interests of developing society? The answer to this question depends to a significant degree on whether we will be able to resolve the problem of managing evolution. The resolution of this problem is unthinkable without knowledge of the patterns of change of a population under natural conditions.

Fundamental principles of evolution were disclosed initially by Charles Darwin, and then, as a result of the synthesis of genetics and Darwinism, were deepened and developed by the science of microevolution formulated in the 1930s and 1940s. But knowledge of principles of evolution in itself is insufficient for a transition to controlled evolution. In order to avoid disturbing the complex natural systems that have developed over millions of years, we must understand what we may do and must not do with natural populations, and we must foresee the effects of various evolutionary factors on them, and know how to find populations under natural conditions and how to study them effectively. All these and many other problems of population biology must be fundamentally resolved from the ecological-evolutionary position in order to understand the change in frequency of occurrence of various traits. The difficulties of genetic studies of existing animal species in nature force scientists to look for special methods to serve this purpose. This is how phenetics had its beginning: it is a genetic approach applied to the study of any natural populations. This approach is based on isolating and counting discrete, alternative, hereditarily determined traits or phenes. On the one hand, phenetics is a part of genetics, and on the other, a part of zoology and botany, ecology and morphology, physiology, ethology, and developmental biology.

Ideas that served as a basis for the present work were formulated in 1969–70 by N. V. Timofeeff-Ressovsky, A. V. Yablokov, and N. V. Glotov in producing *Outline of Population Theory* (1973). It is difficult to name all the people to whom I would like to express appreciation for discussions and constructive criticism of the problems considered in this book. In the first place, I cannot fail to mention my teacher, the late Nikolai Vladimirovich Timoffeef-Ressovsky, who urgently directed my attention to the need for disseminating the evolutionary-genetic approach in the areas of zoology and comparative anatomy in which I was interested. Among many others, I must name N. I. Larina (Saratov), M. V. Mina, M. S. Gilyarov, B. M. Mednikov (Moscow), S. O. Sergievskii (Leningrad), V. N. Bol'shakov (Sverdlovsk), D. K. Belyaev (Novosibirsk), and among my foreign colleagues, R. J. Berry (London), Zd. Puček (Bialowieza, Poland), M. Soule (Los Angeles), W. E. Evans and W. F. Perrin (San Diego), and L. M. Van Valen (Chicago).

Many positions developed in this book were formulated from 1974 to 1978 during lively considerations and discussions in the Laboratory of Postnatal Ontogenesis of the N. K. Kol'tsov Institute of Developmental Biology of the Academy of Sciences of the USSR; at seminars and discussions at the Institute of Paleontology, at the A. N. Severtsov Institute of Evolutionary Morphology and Ecology of Animals, at the Institute of Plant and Animal Ecology of the Academy of Sciences of the USSR, at Kabardino-Balkarsk University in Nalchik, at Kiev University, and at the University of Kirghiz; during my work at the Institute for Mammalian Studies of the Polish Academy of Sciences in Bialowieza and at the Hubbs Sea World Institute in San Diego; and in discussions at Jagellonian University (Krakow), the University of Warsaw, and the University of California at San Diego.

The structure of the book reflects the process that generated the phenetic method. The first and second chapters are a brief presentation of current concepts of microevolution and those problems of evolutionary biology for which the phenetic approach is required. The third chapter is historical and demonstrates the origins of this approach in genetics, botany, zoology, and paleontology. In the last three chapters a detailed analysis of the concept of the "phene" precedes the presentation of the content of phenetics, including an account of the phene pool and phenogeography.

PHENETICS: EVOLUTION, POPULATION, TRAIT

CHAPTER ONE

The Mechanism of Evolution

The principal strategic problem of biology as a scientific discipline is the recognition of patterns of development of life in order to manage them in the interests of man. As early as the 1920s, N. I. Vavilov placed before scientists the problem of moving to managed evolution as the cardinal problem of biology. Managing evolution means constructing organisms with desired characteristics and properties, according to the wishes of man, and creating biogeocenoses that would not be inferior to the natural ones in richness and variety, highly coproductive agrocenoses that would ensure the required amount of food for man, etc.

Of course, biology has always been a basis for solving a multitude of practical problems connected with agriculture, forestry, and medicine. But if in the past problems connected with man's winning a place in the biosphere, with confirming his place in it, and then with subordinating the biosphere completely confronted biology in sequence, today the problem is formulated somewhat differently. Having mastered the biosphere, man must not allow its impoverishment, the destruction even of its separate components (whether these be species with unique gene pools or biogeocenoses with unique population collections of hundreds or thousands of species adapted to one another over millions of years and forming a system that is at once mobile and stable).

The notable impoverishment of the biological component of the noosphere, the sphere of rational human influence on nature, must of necessity have a negative effect on the existence of man on earth. People, biosocial beings, are the product of biological evolution and are an inherent part of the biosphere. Human biology requires the preservation and support of the qualitative diversity of what is living in nature. For this reason in the foreseeable future theoretical problems in biology will be closely linked with the preservation of the environment, the main problem of ecologically developing humanity.

Moving from general discussions to current problems in biological studies, I will try to determine what is known at present about organic evolution. To manage evolutionary processes, obtain new traits and properties, and create new forms of organisms, we must know those interacting phenomena and processes that will permit us to move in the future toward managing the evolution of biological material.

Population: The Elementary Unit of Evolution

Where in nature do the elementary evolutionary processes take place? Formerly it was thought that the arena for these processes was the species itself as a whole. Beginning with the notable work of S. S. Tchetverikov, "Certain Aspects of the Evolutionary Process from the Standpoint of Contemporary Genetics" (1926), and subsequent works of an entire pleiad of geneticists and evolutionists in various countries, primarily R. Fisher, S. Wright, and J. Haldane, the fundamental significance of the fact that a species could be divided into separate groups of individuals became apparent. It became clear that a species has a quite complex structure that can be split into numerous subspecies in some cases, or displays quite complex gradual changes in traits and properties over great spaces. (This does not permit the drawing of sharp subspecies boundaries, but indicates the nonhomogeneity of the species as a whole.)

Investigators in the area of population genetics were the first to understand the fundamental significance of the comparatively small groups of individuals on which the population of a given species depends. It is in just these small groups that all processes of primary exchange of genetic material take place and processes of selection and all other unseen evolutionary changes of living nature occur; this was subsequently called the process of *microevolution*.

There is no species of living organism whose population could have been distributed through space completely uniformly. The inescapable mosaic property of environmental conditions determines, for example, the increased concentration of frogs in marshes and nightingales and warblers in places with a well-developed undergrowth. As a rule, such density centers of a species population are preserved for the duration of many generations of individuals, and it is specifically within these centers that most free interbreeding takes place. Such groups of individuals are called populations. A population is a sufficiently numerous aggregation of individuals of a determined species over a long period of time (a large number of generations) which inhabits a determined area of geographical space within which some degree of chance interbreeding—panmixis—occurs and where no notable isolating barriers exist. It is separated from neighboring similar aggregations of individuals of the given species by some barrier that prevents free interbreeding.

For an understanding of the basic mechanisms of evolutionary processes, a precise, noncontradictory, and adequately complete definition of a population is extremely important, and for this reason I will consider the basic conditions of the definition presented above in greater detail.

By no means can all groups of individuals be called populations in the evolutionary-genetics sense. A population is a group of individuals existing for a long time and stably maintaining itself through many generations. Along the boundaries of the species area, there may from time to time appear short-lived groups of individuals. An example of this is the movement of a flock of Pallas' sand grouse (*Syrrhaptes paradoxus*), small birds of the order Pterocletes that live in the zone of Asiatic deserts and semideserts extending from the lower reaches of the Volga to beyond the Baikal and into Mongolia, far beyond the limits of this area. These birds were observed many times in Europe, and they nested for several years in Denmark, Holland, and in the Ukraine (figure 1.1). But in the end, these small groups, being incapable of adapting to conditions other than those of their native

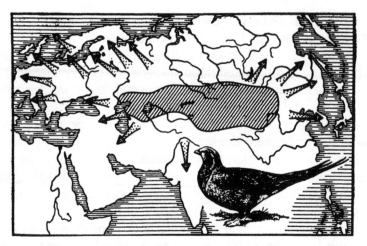

Figure 1.1. Areas of flights of separate flocks of *Syrrhaptes paradoxus* beyond the species range.

habitat, died out everywhere. Such groups of birds cannot, of course, be termed populations.

An important criterion for a population is the degree of free interbreeding which must, of course, be higher within it than between neighboring populations. In different species, the degree of panmixis in the populations may fluctuate significantly. In some species individuals form pairs for life (for example, the swan); in others, only for the mating season (for example, many ducks); in still others, the females are fertile only once in a lifetime (many insects, arachnids). Finally, in some species fertilization takes place externally (most water organisms—for example, fish, amphibians). In fish, which have massive spawning, fertilization of a group of egg cells may be done by a mixture of spermatozoa from various individuals. All these types of interbreeding and methods of fertilization affect the degree of free interbreeding in a population of a given species.

We must, however, emphasize that a high degree of panmixis is not a sufficient basis for recognizing a group of individuals as a population. Thus, in recent years it has become clear that in many species within a population the members can be divided at any given moment into small and comparatively isolated groups of individuals, demes (from the Greek *demos*, nation) (figure 1.2), the degree of pan-

Figure 1.2. Population of bank voles in a forest near Moscow. Disinct groups of animals isolated spatially and phenetically; dots indicate where individual animals were caught; trees indicate territory of old forest. (Krylov and Yablokov 1972)

mixis in which is somewhat higher than in the population as a whole. These small intrapopulation groups cannot be considered as independent formations. They exist only for a short time, usually for one or two generations. Such ephemeral groups of individuals cannot be considered as independent evolutionary units.

Populations within one species or another may occupy different territories and aquatories. Unfortunately, very little precise data exists on the sizes of population habitats. Generally, in species with more mobile individuals, the habitat of the population will be comparatively large, and in species with individuals that are less mobile, it will be smaller. But even here exceptions are possible, of course. Thus, in one of the sections of a park forest approximately 10 hectares in area, there were four or five small areas far removed from one another occupied by separate populations of one species of drosophila (*Drosophila melanogaster*) and only two neighboring populations of another species (*D. obscura*). One population of lizard, *Lacerta agiles*, can live on a territory totaling 0.1 hectare, and another, on a territory of 20 or 30 hectares.

An important characteristic of every population is the number of individuals included in it, which may also be quite variable. For example, populations of some insects include hundreds of thousands of individuals (the same is also true of some small plants). On the other hand, a population of large vertebrates may include only several hundred individuals (the population of a White Sea herd of harp seals is made up of approximately a million individuals, but a population of ringed seals in Lake Ladoga includes not more than 10,000 individuals).

Both large and small populations can be found within each species. Thus, in some cases, *Lacerta agilis* lizards form populations composed of several hundred, and in other cases, of many thousands of individuals.

With very rare exceptions (evidently involving only the very early or the terminal stages of the existence of a species—their inception or dying off), all species consist of many populations, sometimes of very large numbers. Rare species are made up of only several or even of a single population, for example, certain species-endemics, relict, and insular species. The polar bear in the Arctic and the Baikal seal are examples of species of large mammals that are constituted, evidently, of one or only several populations.

Absence of notable isolating barriers within a population, on the one hand, and isolation from neighboring populations, on the other, are usually linked to the degree of individual activity of the members of a population. Mobility of individual animals may be visually expressed in the form of so-called radii of individual activity (this means the distance that a single individual actually negotiates during its life span, or more exactly, the distance between places of birth and breeding). On the average, the radius of individual activity of a duck (*Anas crecca*) consists of hundreds of kilometers; that of a house sparrow or rabbit, approximately 3 km; hare, 5 km; muskrat, 400 m in all; drosophila, 100–300 m; lizards, 30–70 m; and snails (*Cepara*), 8 m. It is clear that populations of muskrat will occupy a comparatively small area, and populations of duck, vast territories.

For plants the radius of individual activity is computed by the distances to which pollen is distributed, vegetative parts (cuttings, buds, runners, tubers, etc.) grow, or seeds are carried (by wind, water, animals). Unfortunately, there is very little precise data of this kind. We know, for example, that pollen from an oak in a forest is distributed only over several hundred meters, and for such herbaceous plants as Senecio, Liatris, and phlox, only over several meters.

Studies of various species under natural conditions demonstrate that populations are always mixtures of various genotypes. Genotypic differences between different populations are expressed even more strongly. The existence and interaction of populations different in genotypic composition promote a complex population structure of a species, which is probably the basis for its relative stability in evolution since the more complex the biological system, the more stable it usually is.

Why is it that the elementary evolutionary unit is specifically the population and not the species, as was thought earlier, or the individual, as was also proposed at times? The separate individual cannot be considered a unit of evolution because it does not embody its own evolutionary fate. This means that the individual cannot exist for a long enough time on the scale of the evolutionary process that continues for thousands and millions of years. For this same reason such short-term associations of individuals as, for example, the family, the herd, or the flock cannot be elementary evolutionary units. The species as a whole contains its own *evolutionary fate*, and can, for this reason, be considered as an evolutionary unit. But every species, as a rule, consists of populations, each of which is capable of existing for an unlimited span of time. For this reason the population, and not the species, plays the role of the elementary evolutionary unit. The significance of the species in evolution is not in the least diminished by this: the species serves as a nodal instant in the evolutionary process, the qualitative stage of evolution.

When it appears, every new species passes a stage when it is only a population (or group of populations) of the parent type, but this does not mean that every population will turn into a new species in the future. This is potentially possible, of course; an indication of this is the evolutionary independence of every population, the possibility of its developing further under specific conditions of existence. But these conditions are not always found in optimal relations with the features of each forming population, and the requisite isolating barriers rarely develop. For this reason few populations develop into independent species.

The Smallest Evolutionary Change

The conclusion as to the possibility and need for designating a population as the elementary evolutionary unit links the evolutionary process with changes taking place within the population. As data from genetics indicate, these changes must affect a specific range of various genotypes. (As a result of the demise of separate individuals, the concentration of some forms, alleles, of various genes may decrease, and the ratio of various genotypes may change.) When external conditions are relatively stable, the genotypic composition of a population, changing slightly, may remain unchanged on the average for a more or less long period. If the influence (pressure) of evolutionary forces on a population (the intensity of action of these forces must be understood as the influence) is sufficiently great or sufficiently long, then the genotypic composition of the population will change for a prolonged period. An evolutionarily significant change will occur in the population, that is, an elementary evolutionary phenomenon.

Thus, the elementary evolutionary phenomenon is a long-term (effective over the lifespan of several generations) and directional change in the genotypic composition of a population. Without such a change, any kind of evolutionary process in a population would be impossible. At the same time, a change in the genotypic composition of a population in itself cannot yet be called an evolutionary process. Such a change in a population as of an elementary evolutionary structure is absolutely indispensable and is a prerequisite for any evolutionary change; without this the process of evolution is impossible. The action of evolutionary factors on elementary evolutionary material is necessary in order for even a minimal evolutionary change to take place in a population.

Evolutionary Material

From what has been said above, it is already clear that elementary evolutionary material consists of changes that stably alter the characteristics of a whole population, that is, hereditary changes in traits. The only known elementary hereditary changes are mutations of various types.

What do we know of the properties of mutation as elementary evolutionary material? We know that mutations arise in all known organisms without exception, and that their frequency of occurrence is quite high: from several percent of the gametes (in bacteria, single-celled algae) to 20–25 percent (in drosophila) per generation carry various mutations. Mutations may involve any morphological, physiological, and etiological traits of the organisms, even such complex traits as relative viability. Examples of mutations have been studied which, arising in a single population, enter the evolutionary arena and then spread to neighboring populations. Some populations close to each other differ from one another by the occurrence, or frequency, of only one or several mutations.

Patterns of occurrence of mutations in populations are the subject matter of population genetics.

The Mutation Process: The First Supplier of Evolutionary Material

Elementary evolutionary factors are isolated on the basis of their character and the nature of their effect on populations and according to the results of the pressure they exert on populations. Here the isolation of four basic *elementary* evolutionary factors seems necessary and adequate. First I will consider the mutation process.

Units of hereditary variability, mutations, form the elementary evolutionary material. But the process of the mutations' appearing is in itself an elementary evolutionary factor that exerts a specific pressure on the genetic structure of a population. What determines the degree of possible pressure of this factor on a population?

The frequency of appearance of separate mutations is always relatively low, 10^{-4} to 10^{-6} per generation. But because of the large number of genes, the total frequency of mutations occurring in living organisms is relatively large. It fluctuates, as has been said, within the limits of several to 20 or 30 percent or more per generation.

Consequently, the mutation process exhibits a very appreciable pressure on a population. It changes the original traits and properties in various directions, resulting in "indeterminate variability" in a classical form. This indeterminacy of the mutation process excludes the possibility of its directing the course of evolutionary changes. Free accumulation of mutations in a population may lead only to disturbing those more complex systems that are represented by the individual, the population, and the species as a whole. Under the influence of mutations, the original organization of individuals is continuously disturbed. One side of the mutation process, one of enormous evolutionary significance, consists of breaking down the old that has outlived its time; the other side consists of providing a continuous supply of material for ever newer changes in a population; this is its most important role as a supplier of elementary evolutionary material in the process of historical development.

Fluctuations in Numbers:
The Second Supplier
of Material for Evolution

Another most important evolutionary factor is the change in numbers of individuals, *population waves*. In this case I am speaking of fluctuations in positive and negative directions that alternate more or less regularly, and not of a stable directed process of increase or decrease in numbers in a population.

In nature no species of animal or plant maintains constant numbers of individuals from year to year and from generation to generation. In all populations changes occur in numbers, sharp in some cases and less noticeable in others. The scale of such changes may be great. For example, in even years the numbers of emerging May beetles in western Siberia may fluctuate by a factor of several million. Similar fluctuations in numbers are known in such massive species of animals as mayflies, mosquitoes, and mouselike rodents (which produce so-called years of mouse invasions).

The mechanism of change in numbers is known and has been explained in some cases, and is less well known in others. Sometimes a decrease in numbers is connected with the development of conditions that inhibit survival—for example, a period of cold weather in our climate for insects, amphibians, and reptiles. In other cases, predators significantly influence the periodic changes in numbers.

Whatever the mechanism of changes in numbers might be, it is clear that the number of individuals in a population may be affected by many factors. These factors unavoidably lead to periodic or nonperiodic, seasonal or daily, annual or multiyear changes in the numbers of reproductive individuals in any of the known species of animals and plants on earth.

The significance of such population waves, or "waves of life," for evolution is very great and was first emphasized by S. S. Tchetverikov (1905), who introduced this term into science. Tchetverikov called attention to the fact that with changes in numbers of individuals in a population, the intensity of natural selection is altered. But this is only one of the evolutionary consequences of the wave of life. Another, more important consequence is the possibility of a sharp change in the genotypic structure of a population and the bringing out of formerly rare mutations into a different environment. This is a unique kind of approbation for rare genotypes, a determination of the degree of their viability.

Population waves, like the mutation process, are a factor-supplier of evolutionary material that carries a number of genotypes into the evolutionary arena accidentally and without direction. The pressure of this factor may be quite various and probably normally exceeds that of the mutation process.

Isolation:
A Reinforcing Factor
of Differences in Evolution

Isolation, the appearance of barriers that interfere with free interbreeding, or panmixis, is an important elementary evolutionary factor. Destroying panmixis, it fixes accidental differences in complements of genotypes (differences that have arisen as a result of the working of the mutation process and waves of life and due to the influence of selection in various parts of a population). In other words, isolation is a factor in the acceleration and fixing of developing differentiations. As a result of its effect, two or more genotypically different populations are formed from a single original population.

In nature we can find many different cases and forms of isolation, which can be classified quite distinctly. First of all, there are two basic types of isolation: physical (spatial, geographic) in which a population is divided into two or several parts by barriers lying outside it—that is, not connected with biological differences between individuals that make it up—and biological, in which one degree or another of isolation within the population is based on the appearance of corresponding biological differences. Biological isolation can be quite precisely subdivided into three basic forms: ecological-ethological, morphological, and genetic, in *sensu stricto*.

The end result of the effects of all forms of isolation is essentially the same. They elicit and fix group differences by disrupting panmixis, and always lead to leveling the differences caused by interbreeding. Isolation in itself cannot, of course, produce new forms. The presence of genetic qualitative variability is required for this. In other words, isolation, providing the initial stages and reinforcing divergence, always interacts with factors-suppliers of elementary evolutionary material (with the mutation process and with population waves). Regardless of the duration of its effect, isolation cannot be considered a directing factor of evolution; it only promotes the division of the original population and intensifies it.

Isolating barriers may vary, and may range from having little effect to complete isolation. The effect of isolation in most cases, as with most population waves, evidently exceeds the pressure of the mutation process.

Natural Selection:
The Only Directional Factor
in Evolution

Natural selection is undoubtedly the most important evolutionary factor. In describing it, Charles Darwin used the concept "survival of the fittest." What is implied, of course, is not simply survival, but survival and subsequent reproduction; it is just here that the individual has a basic significance for evolution.

Natural selection is a process directed toward increasing or decreasing the probability of progeny for one form of organism as against another form. The basis for selection is the total relative viability of individuals of a specific genotype at all stages; reaching reproductive age and the possibility of leaving progeny depends on this. Selection is effective primarily within the limits of each population, which selects or rejects one genotype or another included in its composition. Carriers of specific traits or properties, specific individuals or groups of individuals, are the material of selection.

Every natural population always represents a certain mixture of various genotypes. Different genotypes in a population are usually represented in different concentrations and may differ morphologically from each other. Under relatively stable environmental conditions the dominant genotypes will always preserve their dominant position, and all deviations from this group will be eliminated. Such a form of selection is called, "centripetal," or "stabilizing" selection.

When living conditions change, however, selection may occur that leads to a change in the average type in the population, to a replacement of one quantitatively dominant genotype by another. This form of selection is called "driving" or "leading" selection, and is the essence of the classical Darwinian concept of selection.

It should be emphasized that any given group of genotypes, carriers of the basic selected trait or property, is always subject to being either picked up or rejected by selection. In this case a group of so-called modifier genes, that is, genes that change the selective significance of the given trait, either strengthening or weakening the effect of the basic gene, may be formed automatically.

It should be noted that this aggregate of automatically selected genes (when selection for the principal trait is sufficiently strong) may result in spreading traits and properties correlatively linked with the principal selected trait in the population. Diversity of situations occurring in nature is quite high, so that sooner or later these secondary traits may become principal objects of selection.

Effectiveness of selection depends in the first place on its pressure and on the duration of its influence in a given direction. The pressure of selection is that degree of difference in relative viability of competing forms, which can be expressed quantitatively—for example, in percents. Direction of selection means the positive or negative selection of given genotypes.

Selection pressure can theoretically vary from 0 to 100 percent dominance of the selected form per generation. but in reality, absence of selection (its pressure equal to zero) is possible neither in nature nor in an experiment: some genotypes will always vary somewhat in the probability of leaving offspring.

In natural populations, where there is always a heterogeneous mixture of individuals and where processes of selection in various directions and with various pressures are always taking place, cases of selection of not one specific genotype, but several genotypes differing from each other frequently occur. In addition, we must not forget that at any given moment selection does not evaluate the genotype itself, but its external expression under given concrete conditions, the phenotype. When we speak of direction of selection, however, what is understood is not one generation, but a whole succession of generations; in this case the mechanism of change in phenotype may be understood only as the result of a corresponding change in the genotype, the system that determines phenotype.

In nature, pressure of selection usually masks the pressure of the mutation process of population waves. Pressure of isolation only reinforces the effectiveness of selection.

Natural selection is the only (and sufficient) elementary factor directing evolution; its effect is always directed by prevailing conditions of existence.

Interaction
of Evolutionary Forces:
The Mechanism of Evolution

The mechanism that triggers evolution functions as a result of the simultaneous action of evolutionary factors within a population as an evolutionary unit. Every population in any species is subject to some pressure on the part of all the elementary evolutionary factors.

Actually, the mutation process is constantly taking place in all organisms. The numbers of individuals fluctuate in all populations. The specific pressure of isolation enters into determining the concept "population," and natural selection is always present in nature. The effect of these factors can change independently and often very sharply.

The pressure of the mutation process probably changed over various geological eras and periods and will undoubtedly change in our time in connection with sharp, local increases in the level of chemical and physical mutagens (for example, in connection with radioactive contamination of regions, use of strong chemicals, etc.). Historically, the range of fluctuations in numbers has undoubtedly changed in every species, and sharp isolating barriers have appeared or the significance of existing barriers has decreased. Finally, the significance of natural selection constantly changes, depending on the changing environment: it can alter its direction, and its intensity can increase or decrease drastically.

As a result of the action of evolutionary forces in each population, elementary evolutionary changes have appeared thousands of times. Over time some of these are integrated and lead to the appearance of new adaptations, and this is the basis of species formation.

REFERENCES

Krylov, D. G. and A. V. Yablokov. 1972. Epigenetic polymorphism in the skull structure of the bank vole (Clethrionomys glareolus). (In Russian; English summary.) Zool. Zhurn. 51(4):576–584.

Tchetverikov, S. S. 1905. "Waves of Life" (from Observations of Lepidoptera, 1903). (In Russian.) Dnevn"ik Zoolog. Otdeleniya Imperatorskogo Ob-va Lyubitelei Estestvoznaniya, Etnografii (Journal of the Zoological Department of the Imperial Society of Naturalists and Ethnographers) 3(6):103–105.

FOR ADDITIONAL READING

Ayala, F. J. and J. W. Valentine. 1979. Evolving: The Theory and Processes of Organic Evolution. Menlo Park, Calif.: Benjamin Cummings.

Darwin, C. 1859. *The Origin of Species by Means of Natural Selection; or, The Preservation of Favored Races in the Struggle for Life.* New York: Avenel Books, 1979.

Dobzhansky, T. 1951. *Genetics and the Origin of Species.* 3d ed. New York: Columbia University Press.

Dobzhansky, T., F. J. Ayala, G. L. Stebbins, and J. W. Valentine. 1977. *Evolution.* San Francisco: W. H. Freeman.

Fisher, R. A. 1930, *The Genetic Theory of Natural Selection.* Oxford: Clarendon Press.

Grant, Verne. 1977 *Organismic Evolution.* San Francisco: W. H. Freeman.

Grant, Verne. 1980. Gene flow and the homogeneity of species populations. *Biol. Zentralblad.* 99:157–169.

Haldane, J. B. S. 1932. *The Causes of Evolution.* New York: Harper.

Huxley, J. 1974. *Evolution: The Modern Synthesis.* 3d ed. London: Allen and Unwin.

Mayr, E. 1970. *Populations, Species, and Evolution.* Cambridge: Harvard University Press.

Schmalhausen, I. I. 1949. *Factors of Evolution: The Theory of Stabilizing Selection.* (Translated from Russian.) Philadelphia: Blakiston.

Sperlich, D. 1973. *Populations Genetik.* Stuttgart: Fisher.

Tchetverikov, S. S. 1926. On certain aspects of the evolutionary process from the standpoint of genetics (with the author's comments). *Proc. Amer. Philos. Soc.* (1959), 105:167–195.

Timofeeff-Ressovsky, N. V. 1939. Genetics and evolution. *Z. induct Abstammungs und Vererbungslehre* 76(132):158–218.

Timofeeff-Ressovsky, N. V., N. N. Voroncov, and A. V. Yablokov. 1975. *Kurzer Grundriss der Evolutionstheorie.* (Translated from Russian.) Jena: Fisher.

Yablokov, A. V. and A. G. Yusufov. 1981. *Evolutionary Theory,* 2d ed. (In Russian.) Moscow: Vysshaya Shkola.

CHAPTER TWO

The Importance
of the Study of Natural Populations

The preceding chapter was a brief presentation of the bases for the science of microevolution. The fact that this science has been created is a sign of the position attained by evolutionary thought in the first half of the twentieth century. But it should be noted that this science is still at the level of theoretical description of the original developments and formulations of the initial definitions.

Beginning with the fundamental works of R. A. Fisher (1930), S. Wright (1931), J. B. S. Haldane (1932), T. G. Dobzhansky (1937), N. V. Timofeeff-Ressovsky (1939), and a number of other investigators who founded contemporary evolutionary science, this field has seen no great new discoveries during the last twenty-five to thirty years. Of course, science has not halted its progress: a whole series of facts that have significance for the development of separate chapters of the science of microevolution and population genetics has been disclosed. Thus, a most important achievement was the study of features of protein polymorphism (Lewontin, Hubby 1966) that led to an explanation of the degrees of genetic variability in natural populations. It developed that for an adequately broad spectrum of species studied—from drosophila to man—a very similar level of genetic variability in populations is characteristic. On the average, one-third of all genes have several forms (alleles), that is, they are polymorphic, and each individual is heterozygous (carries different alleles) for approximately 10–20 percent of the sites of specific genes in the chromosome (loci). This conclusion was confirmed in a general form by analogous results obtained as early as the 1930s and 1940s by the analysis of the distribution of separate mutations in natural populations of fruit flies in the drosophilid family that were carefully studied in this respect.

Of interest were the many works in subsequent decades on the intraspecific systematics of various groups of animals that displayed a surprisingly large number of species-doubles in groups that had apparently been studied in detail earlier (E. Mayr 1965). According to N. N. Vorontsov, the number of species of mammals in the fauna of the USSR decreased by approximately 20 percent as a result of combining morphologically similar, but genetically and evolutionarily different, forms under one species name. Extremely common and occupying a significant portion of the territory of the USSR is a species, "common field mouse," which has been subdivided into three independent species as a result of detailed studies. The "malaria mosquito," considered as a single species for the last hundred years, turned out to be a complex of species containing, according to different estimations, from seven to fifteen different evolutionary-genetic forms; some of these species differ externally only in the microstructure of the egg surfaces as seen in a scanning electron microscope.

The unexpected riches of these hidden species in various groups were disclosed by the broad use of karyological analysis in population studies—consideration of the number, character, and common structure of chromosomes (Wahrman, Zahavi 1955; Sharman 1956). But even these studies, remarkable in themselves, did not bring anything new in principle to the concept of microevolution.

I might name several other great and important works connected with the study of interspecific features in various groups of plants and animals, such as the study of the role of polyploidy and species formation in plants (Stebbins 1950), the disclosure of the hybridogenetic character of certain species, which demonstrates a distribution, broader than was expected in the 1930s, of phenomena such as so-called reticular or network evolution (Grant 1971), and the possibility in principle of so-called lateral transfer (transduction) by viruses. But the general conclusion that a theoretical standstill had been reached in knowledge about the microevolutionary process did not change because of these. At first glance, this seems surprising. For simultaneously with the development and formulation of micro-evolutionary concepts, there occurred intensive development of population biology in its various aspects. It would seem that a multitude of data on the study of natural populations would give new and interesting material for the development of evolutionary-genetic concepts. But this did not happen. The reasons for this apparent discrepancy are now becoming clear.

In our time almost no one doubts that, in the final analysis, genetic-evolutionary interpretation helps in understanding the mechanisms of development of separate populations and of whole species and species complexes in nature. Moreover, it is impossible to understand the process of evolution apart from its genetic interpretation.

Meanwhile, without evolutionary elucidation, the study of population ecology, population physiology, or ethology is impoverished and loses its general biological sense. But it was precisely evolutionary content that was not included for a long time in most studies in these areas of population biology. Thus, in the 1930s and 1940s a kind of break developed between evolutionary-genetic and ecological-physiological approaches in population biology.

Now there is every reason for seeking a new synthesis, a synthesis of microevolutionary sceince (or in general, the genetic-evolutionary approach in population biology) with the broad range of population studies in ecology, morphology, physiology, botany, zoology, and other biological sciences connected with the study of populations. The need for such a synthesis is felt both on the part of microevolutionary studies and on the part of "nongenetic" population biology.

I have already spoken about the marked standstill in the study of microevolution during the last decades. One of the main reasons for this, if not the most important reason, is the insufficient amount of data concerning the features of the microevolutionary process in nature. For that matter, what kind of material was it that served as the basis for the contemporary theory of microevolution? It was based on data on species of plants and animals that were thoroughly studied genetically. There are comparatively few such species. In the first place there are eight or ten species of drosophila, an ideal material for population-genetic studies, the group of species that is studied genetically more than all the rest of the organisms inhabiting our planet.

In this group, genetic studies, that is, studies from the point of view of heredity of various traits in interbreeding, were done on no fewer that a billion individual crosses. With respect to other invertebrates, there are genetic data on several species of molluscs; one species of cockroach, silkworm, gypsy moth, meal snout moth, Mediterranean flour moth; some wasps; and several species of mosquitoes, flies, and predatory and plant-eating ladybugs. Among protozoans, genetic studies have been made only of certain infusoria; of lower vertebrates, a number of species of breeding fish and two or three species of tailless amphibians; among birds, of domesticated hens, turkeys, mallard ducks, and to a lesser degree, pheasants, pigeons, quails, parrots, and canaries. Among mammals, the species most studied genetically is the house mouse (more than 500 genes are known), and then the black rat, rabbit, guinea pig, and two species of hamsters. Domesticated mammals have been comparatively well studied—cattle, sheep, goats, horses, cats, dogs—as have been fur animals bred in captivity—fox, mink, blue fox, nutria, and sable. Quite a bit of data has been gathered on human genetics and on genetics of two or three species of apes recently bred in captivity. Approximately 3000 inherited traits are known for *Homo sapiens*.

On the whole, it would not be an understatement to say that from the genetic point of view not more than fifty to sixty species of animals have been studied, and thorough studies with analysis of several hundred genes have been done for only three or four species.

The situation with respect to the genetic study of plants is similar. Such studies of cultivated plants have been more successful (primarily maize, pea, bean, tomato, barley, wheat, oats, rye, rice, clover, turnip, sugar beet, rutabaga, cabbage, mustard, radish, potato, sugar cane, sunflower, cucumber, melon, squash, and certain other species). A number of species of plants, useful for various reasons for experiments, have also been studied genetically; among these are snapdragon, chrysanthemum, tobaccos, sweet pea, primula, evening primrose, black cumin, violet, pansy, and certain other ornamentals. Genetics has recently found its "botanical drosophila": crosses of *Arabidopsis thaliana* are very useful for genetic studies. But on the whole, it is true that plant genetics too is based on the study of only twenty or thirty species.

Of the whole enormous kingdom of bacteria genetic studies have basically been done only on coliform bacteria, mouse typhus, a number of pneumococci and streptococci, and several species of the genus *Hemophilus*. Of the fourth kingdom of contemporary organisms, the fungi, genetic studies have been made of only several species of actinomycetes, yeasts and *Aspergillus*, *Ophistoma*, *Penicillium*, and *Micrococcus*. The great majority of species of living organisms (no fewer than 1,500,000) has not been studied genetically. In some cases this is so because for a number of species it is technically impossible (as long as we cannot breed in captivity such species, let us say, as walruses or whales), and in other cases, because of the long period of alternation of generations and extending crossing experiments over decades (many woody plants, reptiles, etc.).

This is a unique situation. To disclose general patterns of microevolution, we must obtain data on the similarities and differences in the process of microevolution in representatives of all large groups of organisms, or at least for several percent of all living organisms. But thus far we know certain basic genetic characteristics of not more than 250–350 species. It appears that population genetics has been studied to any appreciable degree for fewer than a hundred species. Species that have been studied genetically make up not more than 0.02 percent of the total number of existing species; correspondingly, species studied from the population-genetics perspective amount to roughly 0.007 percent! It is obvious that such an insignificant selection from the total number of species can be completely nonrepresentational.

Thus, establishing the universality of phenomena disclosed for the few genetically studied species and the formation of evolutionary patterns is impossible without a manifold increase in our knowledge of genetic-evolutionary facts. Even with the most optimistic approach and the desire to study genetics, scientists will not be able to achieve the necessary broadening of the spectrum of genetically studied species. It will be possible in the next few decades to double, triple, or at most increase fivefold the number of genetically studied species, but not to increase it by a hundred or a thousand times! But it is precisely a thousandfold increase of knowledge that is required for a true elucidation of the patterns of microevolution.

But there is a way out of this situation: The nongenetic sections of population biology must "put on genetic glasses." Such glasses exist; they consist of the phenetics of natural populations.

REFERENCES

Dobzhansky, T. 1937. What is species? *Scientia* May:280–286.

Dobzhansky, T. 1937. Genetic nature of species differences. *Amer. Natur.* 71:404–420.

Fisher, R. A. 1930. *The Genetic Theory of Natural Selection.* Oxford: Clarendon Press.

Grant, Verne. 1971. *Plant Speciation.* 2d ed. New York: Columbia University Press.

Haldane, J. B. S. 1932. *The Causes of Evolution.* New York: Harper.

Lewontin, R. C. and J. L. Hubby. 1966. A molecular approach to the study of genic heterozygosity in a natural populations. II. Amount of variation and degree of heterozygosity in natural populations of *Drosophila pseudoobscura. Genetics* 54:595–609.

Mayr, E. 1965. *Animal Species and Evolution.* Cambridge: Harvard University Press.

Stebbins, G. L. 1950. *Variation and Evolution in Plants.* New York: Columbia University Press.

Timofeeff-Ressovsky, N. V. 1939. Genetics and evolution. *Z. Induct Abstammungs und Vererbungslehre* 76(132):158–218.

Wright, S. 1931. Evolution in Mendelian populations. *Genetics* 16:97–159.

FOR ADDITIONAL READING

Ayala, F. J. and J. H. Kiger, Jr. 1984. *Modern Genetics,* 2d ed. Menlo Park, Calif.: Benjamin/Cummings.

Frankel, O. H. and M. E. Soule. 1981. *Conservation Evolution.* Cambridge: Cambridge University Press.

Lewontin, R. C. 1974. *The Genetic Basis of Evolutionary Change.* New York: Columbia University Press.

Mather, K. 1974. *Genetical Structure of Populations.* London: Chapman & Hall.

Mettler, L. E. and T. G. Gregg. 1969. *Population Genetics and Evolution.* Englewood Cliffs, N.J.: Prentice Hall.

Pianka, E. R. 1978. *Evolutionary Ecology.* 2d ed. New York: Harper & Row.

Sharman, G. B. 1956. Chromosomes of the common shrew. *Nature* 1977:941–942.

Solbrig, O. T. and D. J. Solbrig. 1979. *Introduction to Population Biology and Evolution.* Reading, Mass.: Addison-Wesley.

Timofeeff-Ressovsky, N. V., A. V. Yablokov, and N. V. Glotov. 1977. *Grundriss der Populationslehre.* (Translated from Russian.) Jena: Fisher.

Wahrman, J. and A. Zahavi. 1955. Cytological contributions to the phylogeny and classification of the rodent genus *Gerbillus. Nature* 175:600–602.

CHAPTER THREE

A Brief History of Phenetics

An old saying has it that everything new is the thoroughly forgotten old. To a significant degree this is why the history of science is not an archive or treasury of dead ideas, but rather a collection of unfinished architectural assemblies. Frequently the buildings in these assemblies were unfinished not because of errors in the plans but because of lack of building material. I believe that something similar also occurred in the area of biology that is connected with the study of natural populations and microevolution: the tools—the scientific method for resolving the seemingly insoluble contradiction between the need to study genetic processes taking place in natural populations and the impossibility, now and in the foreseeable future, of studying genetics of even several percent of all existing species of living organisms—had already been forged by our predecessors. For this reason not much history exists.

The Principle of Unit Characters

Gregor Mendel had a remarkable predecessor who almost discovered the patterns of inheritance of traits thirty-five years before Mendel did. This was Augustin Sageret, a French botanist (1763–1851), the author of notable works on hybridization of cucurbits, producer of new varieties of pears and a number of fruit trees (seed and stone). He was the first in the history of the study of heredity to consider the individual traits of interbreeding plants.

One of Sageret's main experiments was crossing melons of the chateau variety (muskmelon) with cantaloupe. The following traits are characteristic for these melons.

Cantaloupe:—	Chateau:—
Yellow meat	White meat
Yellow seeds	White seeds
Reticulate skin	Smooth skin
Distinct ribs	Ribs barely apparent
Pleasant flavor	Sweet, very acid flavor

"The expected hybrid progeny should have the following average traits: 1) white-yellow meat; 2) pale yellow seeds; 3) light and sparse reticulation; 4) weakly expressed ribs; 5) a sweet and simultaneously sour taste," writes Sageret in his only paper on hybrids, "Observations on the Formation of Hybrids, Varieties, and Variants," published in 1825. "But the situation is quite the opposite" (Mendel, Naugin, and Sageret 1968:64).

The hybrids actually had the following traits:

First hybrid—	Second hybrid—
Yellow meat	Yellowish meat
White seeds	White seeds
Reticulate skin	Smooth skin
Clearly defined ribs	No ribs
Sour taste	Pleasant flavor

"There is a much more clearly expressed distribution of various traits without any blending among them." Thus for the first time, the principle of unit characters was established.

A year before Sageret's work, there appeared a major publication, "Some Remarks on the Supposed Influence of Pollen in Crossbreeding Upon the Color of the Seed Coats of Plants and the Qualities of Their Fruits," (1824) by T. A. Knight (1759–1838), an eminent botanist, founder of the London Society of Horticulture, and its first president. Knight experimented with different varieties of pea, quite deliberately selecting this species for research, as Mendel did later. Knight observed the change in color of maize seeds and flowers in various hybrids during the process of interbreeding. He discovered their stability and invariability, that is, in essence he approached an understanding of the phenomenon of dominance based on discreteness, the discontinuity of individual traits.

Mendel's works of genius, which he carried out over a period of eight years and concluded in 1863, were a natural continuation and conclusion of these and a whole series of other attempts to penetrate the essence of heredity by means of hybridization.

These are the traits that Mendel considered in crossing various pea forms:

> seed coat smooth or wrinkled;
> endosperm yellow, orange, or green;
> seed coat white, gray, brownish gray, or dark chestnut
>   with or without violet spots;
> flowers white or purple;
> form of pod convex or with a constriction;
> color of immature pod green or yellow;
> flowers axillary or terminal;
> stem normal or seven to eight times shorter.

As a result of crossbreeding, Mendel demonstrated that these traits ("stably differing," in his terminology) are inherited as isolated units, occur in various combinations, are discrete, stable, and linked with material structures of some kind that are transmitted through the gamic cells. In this way he discovered the very essence of heredity, demonstrating at the same time the only methodically reliable way of studying it, the method involving isolating and considering discrete alternative features.

After telling about the works of Mendel, histories of genetics usually mention the subsequent neglect of his work and the independent rediscovery of the patterns of heredity in 1900 by Hugo de Vries, Erich Tschermak, and Carl Correns. This is not completely correct, since it neglects the succeeding scientific work in the area of heredity and variability in plants and animals that filled the last three decades of the nineteenth century.

Only four years after the publication of Mendel's paper in the Brunn (the present Brno, Czechoslovakia) *Proceedings of the Amateur Naturalists Society*, in 1869 W. Waagen (1841–1900), the German paleontologist, published his monograph "Morphogenesis in *Ammonites subradiatus*." Considering the development of new forms of ammonites in

Figure 3.1. Gregor Mendel, founder of modern genetics. He made alternative traits the basis for his work.

Jurassic deposits, Waagen introduced a new concept to science: "mutation." The mutation of Waagen was an uneven transition from one systemic form to another, as if there were a sharp change through time. But the principal theoretical discovery of Waagen was isolating the "single phyletic character" in phylogenesis. With paleontological material, Waagen thought that it would be possible to trace, as it were, the movement of this single phyletic character through time.

At the beginning of 1884, following a serious illness, the prelate Gregor Mendel died in the monastery in Brno, having long since left science and devoted himself to social-political activity. He died without having received recognition for his discovery, but several months after his death, Carl von Naegeli (1817–1891) published a book in Munich, *The Mechanical-Physiological Theory of Evolution*. Historians of science now believe that this book advanced ideas permeated with the numerous—for that unhurried time, customarily thorough— multipage writings of Mendel and presenting the results of his experiments: "Every visible trait is found in the ideoplasm in the form of a deposit; for this reason, there are as many kinds of ideoplasms as there are combinations of traits."

This statement is important because what is under discussion are the hereditary deposits as representatives of discrete traits of an individual, and not of cells, organs, or part of the body as had been widely assumed at that time by most investigators (including Charles Darwin and August Weismann). By most, but not by all. The Dutch botanist, Hugo de Vries (1848–1935), who had been interested for a long time in the problem of heredity in species formation, proceeded along the same path as von Naegeli: discrete hereditary units, "pangenes," are the agents of heredity; each of these is respsonsible for specific elementary antagonistic traits in an organism. According to de Vries, the study of the problem of the appearance and disappearance of individual, discrete traits could shed light on the problem of the origin and development of species. The author of the future mutation theory of the origin of species supported these theoretical views by extensive experiments on crossbreeding close plant species that differ clearly in antagonistic traits (color, pubescence, thorniness, etc.).

Thus, although Mendel's work did remain buried in several

hundred copies of the *Proceedings of the Amateur Naturalists Society of Brunn*, the movement of biological thought conditioned by it, which was initiated in the works of Sageret and Knight, continued uninterruptedly.

From Discrete Traits
to the Phene

William Bateson, the English biologist, was one of the people who formulated whole scientific trends and perspicaciously distinguished the "essential from the nonessential." He proposed the term "genetics" in 1906 and was the organizer of the first conferences on hybridization that initiated the series of international genetic congresses (the first conference was held in 1899 in London). It was this William Bateson that N. I. Vavilov called "my teacher."

In 1894, Bateson published his solid report (610 pages of small-type text with 209 illustrations) titled *Materials for the Study of Variation, Treated with Special Regard for Discontinuity in the Origin of Species.* In essence, life is discontinuous, discrete, and the basis for this is discreteness of heredity. Natural selection acting, as proposed by Darwin, on the basis of very small deviations, continuous variability, in some way leads to discontinuity of species. Resolving the problem may come not through the study of adaptation (which will always remain a fascinating part of natural history), but through the study of "discontinuous variations." And thus, Bateson concluded, the principal difficulties with Darwinism in elucidating species formation lie in the transition from continuous variability to discontinuity of species. To resolve this problem it is necessary to get away from the traditional view of variability as a continuous phenomenon: variability may be discrete. The sweeping summary of the data on discontinuous variability of traits in nature set forth by Bateson did not contain any kind of new doctrine; in his opinion, it only "[brought] together materials that will help others, in the future, to continue to resolve this problem" (1894; 567).

With this book, Bateson immortalized his name as the proclaimer of the mutation theory. Five years after it was published, appearing at a conference on hybridization, which was later called the First International Congress on Genetics, with a paper, "Hybridization and Interbreeding as Methods of Scientific Investigation," Bateson was one of the first in modern scientific literature to pose the question of the importance and necessity of considering "unit characters": "At present we are no longer poor with respect to general ideas on evolution. We are in need of detailed knowledge of the evolution of separate forms" (cited from Gaisinovich 1967:127).

Several years after Mendel's laws were rediscovered in 1900, the fundamentals of this science, which today we call genetics, were formulated. But meanwhile, until 1906 when the word, "genetics," came from the from the facile hand of Bateson, this was not yet an independent discipline, but a branch of experimental biology that had to do with the study of inherited variability.

To denote discrete traits almost every investigator proposed his own term: de Vries, 1901, mutation characters, mutation; Sutton, 1903, single characters; T. H. Morgan, 1910, least characters, elementary characters, alternate or elementary variations; Correns, 1902, independent characters; Bateson and Saunders, 1902, allelomorphs, paarling, paired traits; W. E. Castle and P. B. Hadley, 1915, visible Mendelizing characters, Mendelian unit characters; H. F. Osborn, 1915, biocharacters; and so forth.

If we add to this list Darwin's "unique changes," Mendel's "constant traits," de Vries' "antagonistic traits," and Bateson's "discontinuous variations," it will become clear that it was absolutely necessary to unify in some way multiple, but essentially very close or similar, concepts. Such an attempt was made by H. G. Shull, the American geneticist, specifically regarding the problem of usage of various genetic terms in English. He defines "unit character" thus: "alternative differences of any kind which are either absent or present as a whole in each individual and which have the potential to be united in new combinations with other traits" (1915).

This precise and capacious definition was not given deserved attention. In those years, a new terminology and new points of view were spreading, and new directions in research were beginning to develop that were going ever further from the classical ones. The basis for these was laid down by the eminent Danish geneticist W. L.

Johannsen in the book published in 1909, *Elemente der exakten Erblichke-itslehre*, in which the long awaited terms, "gene," "genotype," "phenotype," and "allele" were introduced into science.

According to Johannsen, the gene is a truly existing, independent unit of heredity that combines and separates in crossbreeding, an independently inherited hereditary factor; an aggregate of genes makes up a genotype. The phenotype is an aggregate of all external and internal traits—"it is the expression of very complex interrelations" (cited from Gaisinovich 1967). Alleles are forms of the state of the gene that bring about phenotypic differences and are localized on homologous segments of paired chromosomes.

These would seem to complete the spectrum of all the terms required by geneticists. But how do we designate separate traits of the phenotype that clearly express the discreteness of the hereditary material? Here also a jargon variant of terms already proposed was helpful.

The use of jargon in scientific terminology has always existed, and will probably always exist. Mutation, according to definition, is a discontinuous hereditary change that affects the phenotype in a specific way, but in scientific usage, a mutation is also an individual that carries a mutant trait; that is, a mutation is the mutant and the mutant trait itself that results from the mutation. An allele is the form of the state of the gene, but in scientific usage, it is also an alternative hereditary trait and the individual carrying such a trait. The number of similar examples could easily be multiplied.

Such widely disseminated jargon usage of precise genetic terms that relate to discrete hereditary traits of the phenotype was permissible when population genetics studied hundredths of a percent of all species that inhabit our planet. Now, at the threshold of obtaining population-genetic data on an incomparably larger number of forms (by several orders), we must think about making scientific terminology more precise—all the more so since the necessary word had already been produced, and this word is "phene."

Thus far very few people know that in his work in 1909, together with other terms, Johannsen proposed the term "phene" to designate a "simple" trait. He added that the term must not be understood in the sense that the phenotype is composed of phenes as the

Figure 3.2. W. L. Johannsen, Danish geneticist. He introduced the terms "gene," "genotype," "phene," and "phenotype."

genotype is composed of genes. The phene is simply a genotypically determined trait. But it developed that no attention was paid to the concept of the phene (mentioned with a critical nuance by Johannsen, by the way). Actually geneticists did not particularly need it: all their thoughts were directed to the analysis of the genotype. They felt the analysis of the phenotype at that stage would only impede the discovery of those deep processes that characterize the genotype. The terminology noted earlier persisted for some time in genetic literature, but it was soon squeezed out by the jargon usage of the terms, "allele," "gene," and "mutation."

It is interesting to compare the fate of the two concepts "phene" and "gene." The latter spread more and more widely, and became almost the basic concept of contemporary genetics. This was so because step by step the materialistic character of the gene was confirmed, and studies were made of its structure, features of localization in the chromosome, and so on.

The opposite happened with the concept, "phene." To the extent that manifestations of the gene were studied, ever greater complexity in the relations between genotype and phenotype were revealed. The path from "gene to phene," the area of "hereditary realization," still remains one of the least developed areas of modern biology. Studies have shown that in the process of the development of the organism, every gene affects a multitude of various phenotypic traits. On the other hand, many genes affect one and the same trait.

This vagueness and complexity of genetic determination of separate traits was undoubtedly the main reason that an understanding of the phene as the hereditarily determined character of the organism was not widespread. The gene was on the main line of the development of biology, the phene was left to one side. But it was ignored temporarily and as a term rather than as a principle accounting for discrete hereditary traits. The movement initiated by Sageret, Knight, and Mendel, and given new impetus in the works of classical geneticists Bateson, de Vries, Johannsen, and others, continued steadily.

In 1913–14, Nikolai Ivanovich Vavilov, our fellow countryman who later became one of the most eminent modern geneticists, worked in Merton, England, at the Institute of Horticulture, which was headed by, in his words, "the first apostle of the new science" (genetics), William Bateson. Returning from this "Mecca and Medina of the world of genetics," Vavilov set to work in his own Vavilovian way to create genetics. The results of this creativity are well known to biologists all over the world: the law of homologous series in hereditary variability, the theory of the origin of cultivated plants, the theory of the geographic distribution of genocenters, the creation of botanic-geographic principles of plant selection—and this is a list only of his major discoveries. At the time no other biologist in the world so successfully combined in his work fundamental discoveries with the development of practical ways of apply them.

N. I. Vavilov, member of the Central Executive Committee of the USSR, founder and president of the All-Union Academy of Agricultural Sciences, director of the All-Union Institute of Plant Breeding (VIR), and obviously the most active world traveler among scientists

Figure 3.3 William Bateson and Nikolai Vavilov. Nikolai Vavilov, Russian geneticist, author of the law of homologous series in hereditary variability, the theoretical basis for the phenetic approach, and his teacher, the English geneticist William Bateson (on the right).

of the twentieth century simply had no time to devote to purely theoretical studies.

Vavilov compelled theoretical genetics to be practical, to serve people not in the future, but in the present. Here he met with a difficulty that stood in the way of all population biologists: neither then nor in the foreseeable future was it possible to obtain an adequate amount of genetic data for all the organisms that interested him. With fantastic energy he organized a wide network of genetics stations in the Soviet Union at which hundreds of researchers purposefully studied the genetics of various species and forms of cultivated plants. But even if he had increased the number of these stations ten times, he would not have been able to study genetically the diversity of forms of even cultivated plants, let alone their close wild relatives, the knowledge of which was needed for understanding the process of morphogenesis, which would make progress toward managed evolution possible.

Vavilov himself said the following about the genetic study of wheats and grains to which probably the greatest number of his special genetic and selection studies was devoted: "Usually all genetic studies are carried out . . . without consideration of the whole enormous ecological-geographical diversity that actually represents the wheat species.. . . Vast amounts of data from such countries of original cultivation of wheats as Abyssinia, Afghanistan, India, and the Mediterranean countries, were absent, as a rule, from the work of geneticists and plant breeders, and for this reason all determinations of contemporary wheat genetics must be accepted at this time only as first approximations. Comparative genetics of wheat species is practically undeveloped" (1935:86–87). In the concluding section of this monograph, he noted, "The genetic nature of traits for most species has not been studied at all" (p. 228), although in the same work he wrote that "for no other plant has as much work been done as for the genetics of wheats" (p. 75). Actually a selective bibliography of only the basic world literature on selection and wheat genetics, which includes this work of Vavilov, contains approximately 600 titles!

How did Vavilov resolve the contradiction between insufficient genetic study of the majority of species and the necessity even today of conducting selection work not blindly, but based on a solid genetic foundation? "Systematic study of wheat species with specific species characteristics disclosed a *remarkable parallelism* [my emphasis] in the traits of species that were genetically and geographically separated" (p. 41). Considering this "remarkable parallelism," the genetic study of one form allowed Vavilov to apply the data obtained to other forms that had not been genetically studied. Here the *"qualitative traits are most significant"* (my emphasis). The tables included in this work demonstrate exactly which qualitative traits are under consideration (tables 3.1 and 3.2, figure 3.4).

Let us examine "The Law of Homologous Series in Hereditary Variability," one of the principal theoretical works of Vavilov. This is how the author himself formulates the "law":

> 1. Species and genera, genetically close, are characterized by a similar order of hereditary variability with such a regularity that, knowing a number of forms within the limits of one species, it is possible to predict the finding of parallel forms in other species and genera. The closer genera and linneons (species—author) are genetically in the whole system, the greater is the similarity in the order of their variability.
> 2. Whole families of plants in general are characterized by a specific cycle of variability passing through all genera and species comprising the family (1967:37–38).

This law (it might be better to call it a rule) is one of the greatest theoretical generalizations in population and general genetics

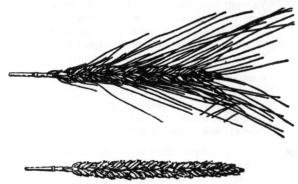

Figure 3.4. Awned and nonawned forms of the wheat *Triticum vulgare*. Presence or absence of awns is an example of the alternative traits used by N. I. Vavilov (1935).

Table 3.1 Certain Traits Used by N. I. Vavilov in the Analysis of Natural Variation in Barley, Wheat, Oats, Wheatgrass, Rye, Maize, Millet, and Other Grains

Trait Group	Individual Traits
Inflorescence	Spikelet body: fragile/nonfragile, simple/branched; Spikelets: awned/nonawned; Awns: coarse/soft, serrated/smooth; Flowers: uniflorous/multiflorous; Bracts: pubescent/glabrous
Grain	Color: white/red/green/black/violet; Surface: membranous/bare; Form: round/elongated; State: glassy/powdery/waxy
Vegetative parts	Leaf: ligulate/nonligulate; glabrous/pubescent; dark green/light green; Culm: hollow/solid; yellow/violet; Sprouts: violet/green; with white bands/without bands; Stalk: glabrous/pubescent; waxy film on stalk, present/absent; Albinism: present/absent
Biological properties	Form of life: winter/summer/semiwinter; Character of flowering: open/closed; Susceptibility to smut, rust, and other parasitic fungi: present/absent.

Table 3.2 Certain Traits Used by N. I. Vavilov to Analyze Natural Variability of Vetch, Pea, Alfalfa, Lentils, Kidney Beans, Fava Beans, Meadow Clover, and Other Plants of the Legume Family

Trait Group	Individual Traits
Flower	Color: white/pink/red/violet/yellow/mottled
Fruit (pod)	Wall: with a parchment layer/without a parchment layer; Form: linear-rhombic/falciform/ensiform/moniliform; Color of immature bean: yellow/green/violet/brown; Color of ripe bean: yellow-green/black/mottled; Surface: pubescent/bare and smooth/convex, bulging/flat
Seeds	Form: spherical/oval/cylindrical/disclike/angular/reniform; Surface: smooth/wrinkled; Pattern: marbled/punctate/spotted; Color of cotyledons: green/yellow/red; Color of hilum: white/brown/black
Vegetative parts	Leaf: with barbels/without barbels, pubescent/glabrous, smooth-edged/serrated, linear/cuneate/oval, yellow/green; Stipules: green/with anthocyan; Stalk: straight/climbing, cylindrical/four-sided/fasciated; Bush: straight-standing/recumbent

in the twentieth century. It serves as the basis for a methodologically bold and innovative procedure: treating variability without genetically studying the forms in concepts and terms of population and evolutionary genetics. Methodologically this procedure is supported by, more precisely based on, analysis in the first place of qualitative, alternative traits, the same traits that have been traditionally used for genetic studies.

If we were to review all of Vavilov's work, we would find these universal examples everywhere: consideration of the variability of qualitative, alternative traits leading to elucidation of the general system of hereditary variability, and the bold, broad comparison of a very limited number of genetically poorly studied forms with forms that had not been studied genetically at all. We also see this in the remarkable *Centers of Origin of Cultivated Plants*, (1926), and in other works.

Nowhere did Vavilov mention phenes; he spoke very little and only casually about individual traits, but all his factual data, extensive tables, and comparisons were made, for the most part, on the basis of comparing easily isolated, hereditarily dependent, discrete traits of the individual. The tools forged by generations of geneticists and consisting of consideration of discrete, alternative traits passed on into reliable hands. It would be more precise to say that Vavilov's genius discovered these tools in an already overflowing store of genetic ideas and methods.

Studying the phenotypes of individuals in natural populations, Vavilov approached them with the only correct genetic measure. Such a measure was the discreteness, the alternativeness, and the qualitative character of traits plus the confidence that if the awnedness of a spikelet, the serratedness of the awn, or the fragility of the spikelet shaft are inherited in one species according to Mendelian laws, then they must behave in the same way in all other close (and not especially close) species.

In the years when Vavilov was writing his outstanding work *Centers of the Origin of Cultivated Plants* (1926) and "Geographical Patterns in the Distribution of Genes in Cultivated Plants" (1927), another eminent and very interesting geneticist, A. S. Serebrovskii, was working in Moscow. Studying the incidence of various hereditary traits in domestic animals (cattle and smaller horned livestock and chickens) throughout vast territories of the Soviet Union, he developed the concept of a gene pool and genogeography. "Casting an eye over our country, we see, for example, a remarkable zonal distribution of coloration of cattle: black-mottled, black, reddish, gray, Ukrainian. We see the same zonal distribution in sheep tails: short, long, thin, fat, straight, and curled, and in a whole series of other traits" (Serebrovskii 1979:72).

In this book, the author turns more than once to concepts of the gene pool and genogeography. Here it is necessary to emphasize the similarity in principle between the works of Vavilov on distribution of discrete hereditary traits in plants and the works of Serebrovskii on distribution in farm animals of the same kinds of traits; Serebrovskii called them "genes," although no real genetic analysis was made for most of the material.

Serebrovskii was the first to propose original methods of vector diagrams for graphic comparison of various populations according to a complex of traits on the model of the phenogeography of chickens (*Gallus domestica*) of Dagestan (North Caucasus).

Vavilov and Serebrovskii differed in their attempts to analyze variability of traits in natural populations from genetic positions. Representatives of the subsequently celebrated Kol'tsoff school of biologists, the students and successors of Tchetverikov, collected and analyzed extensive data from natural drosophila populations. Timofeeff-Ressovsky called the phenomenon of coincidence of phenotypic variability in comparatively few directions specific, and common to related species, "phenotypic parallelism." Published in 1927, the work of Timofeeff-Ressovsky was the beginning of a whole series of studies both in our country and abroad, leading—on the basis of studies of not only drosophila, but also a whole series of other animals—to S. R. Tsarapkin's formulation of the entire course of phenoanalysis at the beginning of the 1940s.

The History of Phenetics

Only comparatively recently have phenes been seriously considered as hereditarily determined traits of the phenotype. Serebrovskii was the first to return to this concept in a philosophical analysis of the problems of evolutionary genetics in 1939–1941 in the book *Certain Problems of Organic Evolution*, not published until 1973. He also used this term when he considered the relationship of the gene and the trait in 1948 in the book *Genetic Analysis* (published in 1970). Beginning in the 1950s, A. Gustaffson, the eminent Swedish botanist-geneticist, used the phene concept more than once in his work.

But this was the situation only formally. Actually discrete traits of the phenotype were being studied from various genetic-evolutionary and taxonomic positions ever more intensively. Especially successful in this were anthropologists who studied the distribution in human populations of hereditarily determined variants of hair and tooth structure, finger and palm prints, and a whole series of other traits. It seems that, in general, anthropologists were the first to think of using discrete phenotypic traits for characterizing separate groups of individuals. In 1893, K. Chembellen, studying variations in sutural bones of the human skull, proposed using their frequencies as a specific populational trait. As early as 1900, the first conclusions from anthropological data on discrete variations of the skull appeared. In later anthropological works (Le Double, Wood-Johnson, Logline, and many others; see Brothwell 1958 and Sjovold 1977) this approach was developed further, and in our time it was continued with a series of elegant works by the English scientists A. C. and R. J. Berry, who indicated the broad possibilities of using phenetic methods for elucidating the origin of ethnic groups and populations and determining the rates of evolution of certain traits.

Paleontologists also found it necessary to isolate and consider phenes in fossil material. As early as 1912–17, the eminent American paleontologist H. F. Osborn made urgent appeals on several occasions for the study of "individual traits " (1915). In this connection, the prominent contemporary Russian paleontologist-evolutionist A. A. Borisyak wrote: "Consideration of every trait separately and its behavior in a series of varying forms would soon lead to an understanding of its significance in the history of the organism. . . . This does not mean a return to the mosaic theory, but only has the purpose of using the method of *discrete traits* [my emphasis] that has proven itself in practice in genetics and other sciences" (1973:73). These appeals were prophetic. Now paleontologists have in their hands hundreds of times more various and abundant data, particularly on invertebrates, that are important for determining the precise geological age of fossil material (by the same token, this is important for stratigraphy and applied geology). For comprehension and deep analysis of this material, paleontologists had to consider discrete traits.

Concurrently with the works mentioned above, data on discrete traits in natural populations were accumulating in population genetics. The frequency and distribution of separate mutations in natural populations always served as a basis for population genetics. In this area hundreds of most interesting works were done based on calculations of the occurrence of phenes, which were frequently called "mutations." Actually, such mutations cannot be analyzed genetically. They can be called mutations only by analogy to phenotypically similar traits studied in the laboratory.

"Ecological genetics," founded in the 1950s by the English geneticist E. B. Ford, was an important step toward joining the genetic approach to zoology and botany, ecology and biogeography. The development of microevolutionary science made this penetration inevitable. At the end of the 1950s, the school of the eminent English geneticist H. Gruneberg developed the study of so-called epigenetic polymorphism, which immediately united geneticists with field population biologists (R. J. Berry 1963, and many others). Only in 1973, however, was a formulation of the subject matter, purposes, and methods of phenetics proposed as a new direction in biology lying at the juncture between genetics, on the one hand, and classical zoology and botany on the other (Timofeeff-Ressovsky and Yablokov 1973; Timofeeff-Ressovsky, Yablokov, and Glotov 1973:278)

Terminology

The relationship between phenetics and others divisions of population biology is presented schematically in figure, 3.5. The diagram shows only the population divisions of various biological disciplines developed quite variously at the present time. Population genetics and

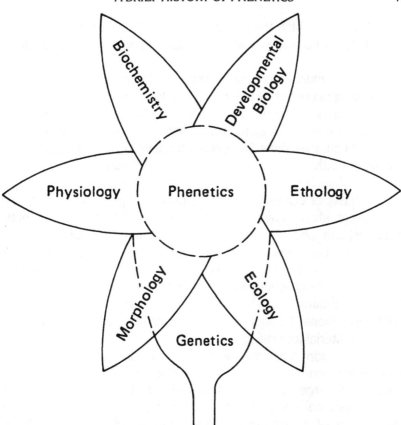

Figure 3.5. Diagram of the relationship of population phenetics and other divisions of population biology (Yablokov 1981).

ecology are evidently the most developed, and in recent years population morphology and ethology have been developing rapidly. Significantly underdeveloped with respect to populational considerations are physiology, biochemistry, and developmental biology.

It seems to me that from this diagram it is quite clear that arguments about the boundaries, for example, between population morphology and phenetics are to a significant degree groundless: in many cases research can refer equally to both population morphology and to phenetics proper. All research can be considered to be phenetic in which the investigator is dealing with frequencies of discrete traits, no matter to which division of population biology these traits pertain.

I must emphasize the substantial difference between understanding phenetics as the genetic-evolutionary approach to studying a population and using "phenetic" terminology in a number of divisons of theoretical systematics. Since the mid-1960s, in English-speaking countries there has been a tendency to call all approaches connected with the use of numerical methods in taxonomy "phenetic." The use of such concepts as "phenetic classification" (classification on the basis of external marks of similarity), "phenetic groups," the isolation of which is also the goal of numerical taxonomy, "phenetic relations," and "phenetic similarity," the determination of which is made by the principal methods of numerical taxonomy, is characteristic for numerical taxonomy. Finally, specific concepts of quantitative taxonomy include the concepts "phenone," a group of forms of one level of phenetic similarity (Sokal and Sneath, 1963, and many others), and OTU (Operational Taxonomic Unit), a group of individuals isolated as primary material for classification.

From what has been said, it is clear that the basis for numerical taxonomy is the task of creating a reliable classification of species and larger taxa that is practical. But the goal of phenetics is the populational study of species under natural conditions to solve the problem of microevolution. This is evidence of the deep and basic differences between these directions in the study of phenotypes. Unfortunately, proposing the term "phenetics" for a new frontier of microevolutionary studies, we (Timoffeef-Ressovsky and Yablokov, 1973) did not take into account the fact that E. Mayr in 1965 defined the content of the whole movement of numerical taxonomy by the term "numerical phenetics." Despite the fact that the terminology proposed by Mayr did not persist, in the years immediately following, in taxonomic literature and in descriptions of interspecific variability, all the morphological traits used for analysis were at times called "phenetics."

It would be incorrect—even useless!—to decree the usage of one term or another in science: only the natural course of events will determine which terminology will be most viable. In this case, for the sake of purity of terminology our direction in research might possibly be designated "population phenetics," and not simply "phenetics."

Thus, in the arsenal of biology there are already adequately developed and approved methods for a study of natural populations of the most varied species of plants and animals before which the researcher dons "genetic eyeglasses," so to speak. The phenetic approach makes it possible to resolve the contradiction between the urgent need for genetic research on an enormous number of species that have not yet been touched by such research and the impossibility either now or in the foreseeable future of actually making a genetic study of these forms.

REFERENCES

Bateson, W. 1894. *Materials for the Study of Variation, Treated with Speical Regard for Discontinuity in the Origin of Species*. London: Macmillan.

Berry, R. J. 1963. Epigenetic polymorphism in wild populations of Mus *musculus. Genetic. Res.* 4:193–299.

Borisyak, A. A. 1973. *Fundamental Problems in Evolutionary Paleontology*. (In Russian.) Moscow, Leningrad: Izd-vo Akad. Nauk SSSR.

Brothwell, D. R. 1958. The use of nonmetrical characters of the skull in differentiating populations. *Deut. Gesell. Anthrop. Ber. über Tagung* 6:103–109.

Gaisinovich, A. E. 1967. *Origin of Genetics*. (In Russian.) Moscow: Nauka.

Johannsen, W. 1909. *Elemente der exakten Erblichkeitslehre*. Jena: Fisher.

Knight, T. A. 1824. Some remarks on the supposed influence of pollen in crossbreeding upon the color of the seed-coats of plants and the qualities of their fruit. *Trans. Hort. Soc.* vol. 5.

Mendel, G. 1865. Versuche über Pflanzenhybriden. *Verh. naturforsch. Vereines Brunn* (1866), 4:3–47.

Mendel, G., C. Naugen, and A. Sageret. 1968. *Collected Works*. (In Russian.) Moscow: Meditsina.

Serebrovskii, A. S. 1929. Problems and methods of genogeography. *Proceedings of the All-Union Genetics Congress* 3:71:74.

Serebrovskii, A. S. 1939. *Certain Problems of Organic Evolution*. (In Russian.) Moscow: Nauka, 1973.

Serebrovskii, A. S. 1940. *Genetic Analysis*. (In Russian.) Moscow: Nauka, 1970.

Shull, H. G. 1915. Genetic definitions in the New Standard Dictionary. *Amer. Natur.* 49(577):59.

Sjovold, T. 1977. Nonmetrical divergence between skeletal populations. *Ossa*, 4, suppl. 1:1–133.

Sokal, R. R. and P. H. A. Sneath. 1963. *Principles of Numerical Taxonomy*. San Francisco and London: Freeman.

Timofeeff-Ressovsky, N. V. and A. V. Yablokov. 1973. Phenes, phenetics, and evolutionary biology. (In Russian.) *Priroda* (Moscow), 5:40–51.

Timofeeff-Ressovsky, N. V., A. V. Yablokov, and N. V. Glotov. 1973. *Outline of Population Science.* (In Russian.) Moscow: Nauka.

Vavilov, N. I. 1926. *Centers of the Origin of Cultivated Plants.* (In Russian and English.) *Trudy po Prikladnoi Botanike i Selektsii* (Leningrad), 16(2):1–248.

Vavilov, N. I. 1927. Geographical patterns in the distribution of genes in cultivated plants. (In Russian and English.) *Trudy po Prikladnoi Botanike, Genetike i Selektsii* 17(3):411–428.

Vavilov, N. I. 1935. The scientific basis for selection in wheat. *Collected Works.* (In Russian.) 2:41–228.

Vavilov, N. I. 1967. The law of homologous series in hereditary variability. *Collected Works*, vol. 1. (In Russian.) Leningrad: Nauka.

Waagen, W. 1869. Die Formenreihe des *Ammonites subradiatus*. *Benecke's Geognost. Palaeontol. Beitrage* 2:179–259.

Yablokov, A. V. 1981. *Evolutionary Theory*, 2d ed. (In Russian.) Moscow: Vysshaya Shkola.

FOR ADDITIONAL READING

Berry, A. C. and R. J. Berry. 1972. Origin and relationships of the ancient Egyptians. *J. Hum. Evolution* 1:199–208.

Castle, W. E. and P. B. Hadley. 1915. The English rabbit and the question of Mendelian unit-character constancy. *Amer. Natur.* 49(577):22–27.

Ford, E. B. 1971. *Ecological Genetics*, 3d ed. London: Methuen.

Gruneberg, H. 1952. The genetics of the mouse. *Bibl. Genet.* 15:650.

Mayr, E. 1965. Numerical phenetics and taxonomic theory. *Syst. Zool.* 14:73–97.

Morgan, T. H. 1910. Chromosomes and heredity. *Amer. Natur.* 44:449.

Osborn, H. F. 1915. Origin of single characters as observed in fossil and living plants. *Amer. Natur.* 49(580):193–239.

Serebrovskii, A. S. 1928. Genogeography and the gene pool of farm animals in the USSR. (In Russian.) *Nauchnoe Slovo* 9:3–22.

Timofeeff-Ressovsky, N. V. and S. R. Zarapkin. 1932. Zur Analyse der Formvariationen. *Biol. Zbl.* 52(3):138–147.

Vavilov, N. I. 1922. The law of homologous series in variation. *J. Genet.* 12(1):47–89.

Yablokov, A. V. 1978. History, current state, and paths of development of phenetic research. In *Physiological and Population Ecology of Animals* 5(7):5–12. (In Russian.) Saratov: Saratov University Press.

CHAPTER FOUR

The Study of Phenes

It has already been said that phenetics is an extension of genetic approaches and principles to species and forms that are difficult or impossible to study. The subject of phenetics is intraspecific variability that leads finally to a consideration of discrete alternative traits of the individual, a consideration of phenes. Methods of phenetics consist of isolating various phenes that are characteristic for the variability of the forms studied and quantitatively and qualitatively studying them in populations and other intra- and interspecific groups of individuals. The goal of phenetics is the resolution of problems of microevolution, theoretical systematics, and practical biotechnical issues connected with the populational study of species under natural conditions.

To provide an initial acquaintance with and to stimulate further development of this new field in population biology, I will consider sequentially only the basic problems arising in the study of phenes as such, the phene pool, and phenogeography, the three basic areas of phenetics. Let me begin by describing phenes.

How Does the Phene Look?

The phene is a separate, discrete, hereditarily determined trait of the individual. This formulation of the concept "phene" is not complete; a more complete definition will be given later after a consideration of a number of examples.

All the mutant characteristics or mutations that geneticists study should be regarded as phenes. The discrete traits of the pea that were selected by Mendel for analysis—smooth or wrinkled form of seeds, yellow or green cotyledons, gray-brown or white color of seed coat, bulging or constricted pod, yellow or green coloration of the ripe pod, axillary or terminal flowers—are discrete, hereditary, alternative. The presence of these traits serves as the basis for isolating one genotype or another. They separate it phenotypically from other individuals, and if we consider any combination of genotypes (in the case of pea, the variety, subspecies), then according to the frequency of occurrence of these traits, we can also distinguish separate groups of individuals.

Let us consider some of the many hundreds of known hereditary, discrete variations of body structure in drosophila. Many traits are connected with the form of the wing. In a fly with a normal wing structure, the wing is always flat. Among individuals that have been studied and analyzed genetically, mutants were found with wings bent in an arch, curled up or down, and with a whole series of other irregularities. Frequently flies were found with notches in the back edge of the wing, with truncated wing edges, with distinctly smaller wings, or with only rudimentary wings. Some flies had no wings at all. Most of the hereditary variations in drosophila appear in the patterns of wing venation: loss of separate longitudinal or lateral veins, branching, thickening or appearance of additional veins, and so forth (figure 4.1).

Quite a number of hereditary variations pertain to features of the setae (macrochaetes), the location of which is an important taxonomic trait for most small insects. As a result of mutation, individual setae may disappear altogether while others may be doubled; sharply truncated or Y-shaped setae may also appear. With respect to other hereditary variations in drosophila, visual examination discloses many traits connected with body coloration or change in type of antennae, feet, abdominal segments, or eyes.

Genes responsible for the appearance of one trait or another (in studies of several hundred mutations) are located on specific chromosomes, and within the chromosome, on its specific segments. Naturally, the traits coded in homologous chromosomes are most often found together (the so-called genetic linkage groups), and traits that are found on sex chromosomes are "linked" with sex.

The appearance of various alleles of a single gene in a

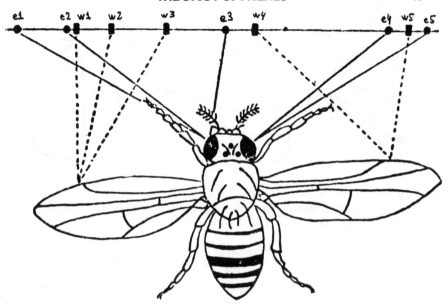

Figure 4.1. Distribution of certain genes that determine the formation of a number of wing and eye structures in the first pair of Drosophila chromosomes. (Auerbach 1961)

phenotype makes it possible to consider the traits of the phenotype that can be called phenes. Let us consider several other examples of phenes whose hereditary character has been sufficiently studied.

The red and black coloration of the two-spotted ladybird beetle is a pair of quite distinct phenes. At the same time, careful genetic studies of one of the species of herbivorous ladybird beetles shows that in this case the angle of inclination of the long axis of the spots on the first wings should be considered a phene (figure 4.2). It is interesting that genetic analysis of a thousand individuals did not disclose any kinds of discrete traits in the phenotype that were connected with the form or size of spots in this species.

In mammals, many phenes may be isolated on the basis of studying the skeleton. Small variations in the form of separate bones, the form of projections, openings for blood vessels and nerves, and the presence of separate additional so-called bregmatic bones in the skull are all hereditarily determined, discrete traits that can be called phenes.

Deviations from the normal structure of the hand or foot,

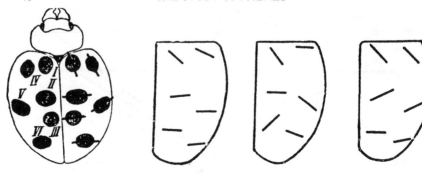

Figure 4.2. Angle of inclination of the long axis of spots on first wings of the herbivorous ladybird beetle *Epilachna chrysomelina*, a discrete, hereditary trait (phene). (Zarapkin and Timofeeff-Ressovsky 1932)

expressed in a decrease or increase in the number of fingers or toes, have been recognized in man for a long time. The appearance of a sixth finger on the hand is connected with a sex-linked mutation. In some families this trait can be traced over many generations.

The hereditary character of the splitting of the fourth or fifth finger in the white dolphin has not been ascertained. But it is quite probable that this is an expression of genotype patterns. Studies have shown that the Far Eastern and northern white dolphin populations differ from each other according to the frequency of occurrence of this phene (figure 4.3).

Undoubtedly, any thinking person can recall many examples of hereditary traits characteristic for his close family and relatives. Such examples may be birthmarks at specific points on the body, the form of eyebrows, nose, lips. A protruding lower lip can be traced in the royal family of the house of Hapsburg for many generations. More than 2800 inherited human traits are now known.

All the examples cited above pertain to nonmeasurable traits, that is, those that cannot be measured. Phenes can also be isolated among measurable traits. The example of dwarf and tall pea plants selected by Mendel for analysis of heredity illustrates this approach very well. In the case of quantitative traits, traits that form two or more discrete variants that do not mask each other may also be phenes—for example, the height of a plant from 10 to 15 cm, from 17 to 25 cm, and from 26 to 37 cm.

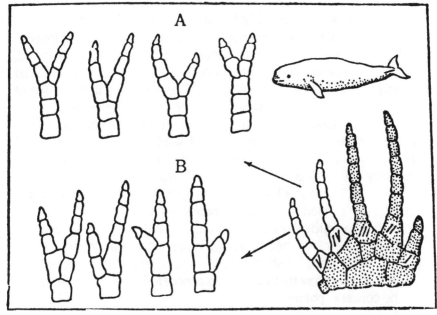

Figure 4.3. Different variations of finger separation in the hand of the white dolphin *Delphinapterus leucas*. The phene of separation of the fourth digit is characteristic for those in the White and Barents Seas; the phene of separation of the fifth finger, for those of the Sea of Okhotsk. (From Timofeeff-Ressovsky et al. 1969)

At present we have accumulated much data pertaining to the study of pure lines of plants and animals. Among these data there are more than a few examples of the existence of discrete variants of quantitative traits. Thus, for example, the average weight of the suprarenal gland in the SZN line of laboratory mice is 44.0 ± 5.7 mg/100 g body weight, but in the C57BL line, it is only 16.0 ± 1.2 mg/100 g (Shire 1976).

Phenes are not only morphological traits of the phenotype, but may pertian to any other traits, both physiological and behavioral. An interesting example of a physiological phene is the insensitivity of rats to a strong poison, the anticoagulant warfarin. This phene was discovered in the 1950s after a successful destruction of rats in a number of Western European countries following massive use of warfarin. In the initial stages rats practically disappeared in most regions where the chemical was used. But after several years rats insensitive to warfarin began to spread. In fact, in some places, because of natural genetic variability in natural populations, separate groups of individuals survived that were insensitive to it. As investigators demonstrated, similar insensitivity has a various genetic nature and is determined largely by the action of at least three different genes in the different populations. These three genes, however, are manifested as a single phene, "insensitivity to warfarin." At present, rats insensitive to the poison are widely distributed in all of Western Europe (Drummond 1970). This is a striking example of the distribution of a new adaptive property due to selection in response to human action.

The examples cited above pertain to phenes of animals. But the gap is filled by the tables, presented in the preceding chapter, according to which Vavilov carried out his studies. The great majority of traits in the plant species studied represent typical phenes.

Having finished this preliminary review of some examples of phenes, I can now try to formulate a more fully developed definition of the concept "phene."

Gene and Phene

From the examples presented, it is clear that we apply the term "phene" to a discrete, alternative trait that reflects hereditary (genotypic) characteristics of the individual.

In order to be precise and consistent, I must first discuss the concept "trait." Sometimes it is quite difficult to define precisely the simplest, most frequently used concepts. Logically, all ways in which subjects or phenomena are similar to each other or in which they differ from each other constitute traits. The main thing that distinguishes a phene from other traits is its diagnostic value: according to the presence of one phene or another, we may form a notion of the genotype.

But the situation with respect to determining the phene is more complex than may appear at first glance. In a general form it is quite clear that any trait is, to one degree or another, determined by heredity; no features of the phenotype can appear without being programmed in the genotype. Thus, not only alternative, discrete traits are hereditarily determined, but also weight, height, and proportions of separate parts of the body.

It is, however, completely clear that hereditary dependence of the dimensions of the body differs in some essential way from the hereditary dependence of eye color. Genetic dependence of body dimensions of hogs includes the possibility of forming, under good conditions, a hog weighing 200 kg, and under poor conditions, only 50 kg. This range of possible variants with one and the same genotype demonstrates the range of the reaction norm. Within its range, variability of a trait is controlled by conditions of development. We may say that it is not the trait that is inherited, but the reaction norm, the capability for forming one set of traits or another under specific conditions of development.

Here we come to a more interesting and little studied problem of the interrelation of the gene and the trait, to the most extensive *terra incognita* in modern biology, the problem of realization of heredity.

Of course, the complexity of realizing hereditary information, the work of the genetic code, is so great that it would be difficult to expect unequivocal and completely constant relations between gene and phene. Sometimes it seems that the action of different genes is hidden under an externally uniform phenotype. It is known, for example, that the phenotypic manifestation of the *black* and *ebony* mutations in drosophila is practically identical: flies with a black body result. Only genetic analysis makes it possible to distinguish these mutations located on different chromosomes. Several different mutations may elicit tailless house mice. With external analysis of the phenotype, the investigator can note only one phene of taillessness. This possibility of masking of different genes by one and the same phene must always be taken into account.

The principles of polymery (several genes affect every trait) and pleiotropy (every gene affects several traits) have been recognized since the beginning of this century (Plate 1910).* This is the phenomenology of the whole process, and the end result is observable. The

* Polypheny is one of the synonyms for the term, pleiotropy; it was proposed as early as 1925 (Haecker 1925); polytopy is another (Timofeeff-Ressovsky 1931).

action of these principles in the development of the individual is presented schematically in figure 4.4. Quite a number of such diagrams, even for species of plants and animals incompletely studied genetically, could be drawn up as long as fifty years ago, and in our time, there could be hundreds.

I will examine in detail only one example of pleiotropy connected with the gene of dwarfism in mice (Berry 1968). This gene is recessive and determines the development of animals that are two or three times smaller than normal; it appeared in one of the progeny of white mice in 1929. Even before the dwarf mice stopped growing, it was possible to identify them according to a whole series of traits such as blunt noses, short ears and tails, flabbiness, timidity, and sensitivity to temperature changes. The life span of dwarf mice is shorter than that of other mice, and both males and females are sterile.

This clear example of pleiotropy in gene action is interesting in one respect. As it happened, dwarf mice could be converted to normal if pieces of rat hypophysis were implanted into their bodies (subcutaneously). The hypophysis is a small gland located at the base of the brain; it regulates the secretion of practically all hormones. After a number of such operations, dwarf animals were transformed:

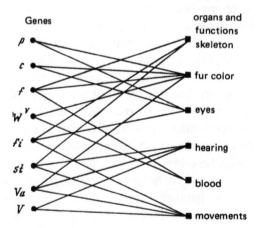

Figure 4.4 Diagram of relation of genes and traits in the house mouse: p = red color; c = albinism; f = bent tail; wv = dominant mottling; fi = nervous behavior; st = trembling; va = rolling walk; v = waltzing movement. (Auerbach 1961)

they attained normal dimensions and began behaving like normal mice, although females remained sterile.

This example demonstrates how traits determined by heredity may be sharply altered by environment or external conditions. This means, to be more precise, that it is not the allele of dwarfism that leads to sterility, early cessation of growth, resistance to starvation, and increased sensitivity to cold, but rather that the action of the dwarfism allele leads to this change in the reaction norm under which, in normal conditions, sterility develops, growth ceases early, and resistance to hunger and sensitivity to changes in temperature increase.

The work of the English geneticist H. Gruneberg, carried out in the 1950s and 1960s, has great significance for understanding the nature of phenes. This scientist compared various mouse lines (maximally uniform genetically as a result of interbreeding) according to the occurrence of minor variants in skeletal structure. These variants included the development of any kind of projections on the bones, the passage of a group of blood vessels in a given location in the skeleton through one large or several small openings, the location of certain openings for passage of blood vessels and nerve trunks, different anomalies in the dental system, and so on.

It turned out that all the variants were of the threshold variety. In the process of embryonal development, as soon as one rudiment or another of a future structure reached a certain size, the formation of the future structure was realized: the bone projection appeared, the large opening was divided by a partition into two, etc. But if the rudiment of the structure did not attain this size in embryogenesis, such traits did not appear in the adult organism. Although the size of the rudiment was affected by many interacting genes and external factors, the decisive factor had to be genetic. The mouse lines studied differed clearly in frequency of such phenes, and these differences were hereditarily transmitted.

Now, after a more detailed consideration of the connection between gene and phene, it becomes clear that the diagram of the relationship between genes and traits presented in figure 4.4 must be made more exact. These relationships are most complex: pleiotropy and polymery are not a property of the genes themselves, but are the result of various reactions directed by the genes, from the simpler, biochemical to the more complex, morphogenetic (figure 4.5). Phenes are visible links of such reactions which can be seen by the observer.

Thus, it would be wrong to consider every phene unequivocally and firmly connected with a single, specific gene. It is probably not the exception, but the rule, that every phene marks one of the alleles of several different genes. This circumstance undoubtedly makes phenes quite broad rather than narrow genotype markers. The presence of one phene or another may be evidence not of the presence of some specific allele of a single gene, but of the presence of one of the alleles of several genes. Having lost in the precision of marking, phenes win in the extent of coverage of the genotype. This is important if we consider that the genotype of higher animals includes several tens of thousands of genes. The consideration of one, ten, or even twenty genes contributes very little to the analysis of population specifics. But the analysis of several hundred genes will be significantly more representative. As noted by R. J. Berry, if one phene is connected with ten or twenty genes, then by studying the frequency of two hundred phenes in a population, we can hope to take into account deviations in the structure of several thousand genes! It is precisely this broadness of coverage of the genotype by phenes that is one of the advantages of studying the morphological, physiological, and ethological phenes before studying the biochemical phenes-electromorphs,

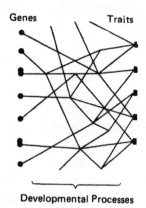

Genes   Traits

Developmental Processes

Figure 4.5. A more precise diagram of the interrelations between the gene and the phene. (Auerbach 1961)

as the results of electrophoretic studies of proteins are sometimes called. Biochemical phenes are, of course, much closer to the hereditary apparatus, the genotype, than the strictly biological phenes, and, as a rule, mark single genes. But even in the best contemporary studies, we can rarely analyze more than several dozen biochemical phenes.

On the whole, the danger of polypheny, which makes genetic marking by phenes at the level of ontogenesis less precise, becomes an advantage in the phenetic approach at the level of the gene pool of a population.

In converting hereditary information from gene to phene, other complexities besides polypheny arise for phenetics. I am speaking primarily of a phenomenon that may be called "ontogenetic noise"—the formation of pseudophenes or "false" phenes, that is, discrete phenotype traits which cannot mark the given genotype. An example of this is the appearance of the phene "innate luxation of the femur" in man. Figure 4.6 shows that a single group of four genes affects the development of the femur joint (the head of the femur and the joint socket in the pelvis); another group of four genes affects the development of the muscles and ligaments. A change in any of these eight genes is reflected in the phenotype as the trait "innate luxation of the femur," a typical phene, according to the frequency of which separate human populations are distinguished. But at least three groups of external factors unrelated to innate characteristics of the given genotype lead to this phenotypic result, the development of a dislocated femur at birth: high estrogen level in the blood of the mother, birth order, and position of the fetus in the uterus. The first of these factors affects only newborn females, while the other two affect both sexes. To this I might add that improper swaddling in the first days after birth also leads to dislocation of the femur.

Clearly, the appearance of a dislocated femur as a result of the action of external factors cannot be a marker of the genotypic composition of a population. This kind of indeterminacy in isolating true phenes, which is always present, is the price we must pay nature when we refuse to study every trait genetically, using the methods of interbreeding over a succession of generations. But we must not exaggerate the significance of this indeterminacy, the level of ontogenetic noise, in the manifestation of phenes. Evidently the complexity

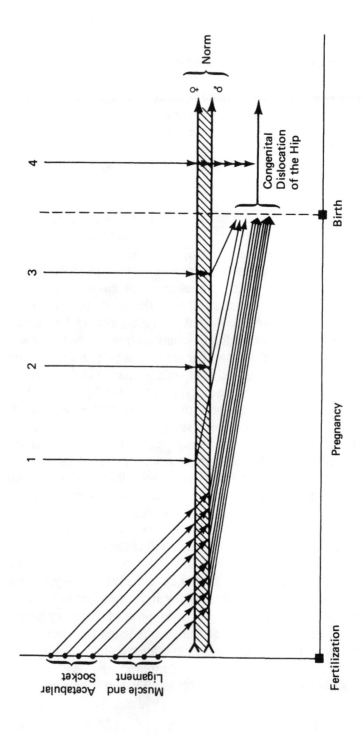

Figure 4.6. Hereditary factors that disrupt normal development: the development of the phene "congenital dislocation of the hip" in the process of human development. (Berry and Berry 1971)

of interaction of external factors in the process of ontogenesis is so great that these factors themselves must in some way be "Brownianized"; that is, they must conduct themselves like particles in Brownian movement and mutually cancel out one another's effects. This conclusion was demonstrated by T. Sjovold (1977) as distinct from its opposite: if it were otherwise, no stable development would ever be observed such as takes place from generation to generation in all species of living beings (as a result of which, in man, even the location of moles, gesture patterns, and voice timber are like those of the parents). But the principal and practical possibility of the appearance of false phenes can never be ignored by the investigator.

On the basis of all that has been said above, I can now present a more complete formulation of the concept "phene." *Phenes are any discrete alternative variations of traits and properties of living organisms that in all available material* (necessarily numerous) *cannot be divided further without losing their quality. Phenes always reflect specific traits of the genetic constitution of a given individual, and in their frequency, the genetic structure of the population and other* (more or less large) *groups of individuals of a given species.*

How Are Phenes Found?

To analyze a trait in genetics, one must only crossbreed a pair of individuals and determine precisely the patterns of any trait or property according to the character of distribution of the trait in the progeny over a series of generations. For phenetic analysis, a large number of individuals is needed. Without such a number, the investigator cannot form a sufficiently complete idea of the discrete variations, which include phenes, that exist under natural conditions.

Experience gained in studying phenes in various groups of plants and animals is the basis for proposing the following way of distinguishing phenes in natural populations:

1. Making comparisons of the observed variability according to separate complexes of traits or properties (coloring, mark-

ings, form of separate parts of the body, variants in structure of different organs and systems, etc.).

2. Distinguishing traits subject to changes due to age and sex. These are complex for a first analysis; although it is possible to find phenes among them, this is a more specialized task. Usually such traits are not included in phenetic studies of animals, although they can be used with great success with plants, as was mentioned earlier.

3. Distinguishing discrete characteristics from total variability of traits and properties, and analyzing them for further divisibility (reduction, subdivision).

Such features of marking as pattern and color should be studied. The discreteness in variability must be found specifically in the color range and not in microscopically determined characteristics of the pigment or microstructure of the surface of the organ that is responsible for the color effect (as in lepidoptera). In the same way, for example, it would be improper to make the pattern of the coloring (bands, spots on separate parts of the body, etc.) a study of the histology of the integument. In both cases the investigator loses the quality of the group of traits being studied and makes a transition to the study of an altogether different group of traits and properties, among which qualitative, alternative, discrete traits may also be found.

4. Analyzing data on genetics of phylogenetically close forms. In order to examine the coloring of the bank vole, it is necessary to become familiar with work in the genetics of coloring of other rodents (house mouse, rat, etc.). Knowing the law of homologous series in the heredity of variability, we may assume that general patterns of coloring in mammals will have some common essentials. Six basic genes, similar in various species of mammals, are responsible for integument color, as are several similar genes for pattern. Of course, in each species several basic genes may be complemented by others, and their action may be modified to various degrees, but the very fact of the presence of homologous genetic bases makes it possible to be confident in making a phenetic analysis of the coloring of practically all mammals.

5 Analyzing indirect data on the hereditary character of separate traits in the given species (expression of traits in parents and offspring, etc.).

6 Confirming the correctness of isolating phenes in natural material. Usually correctly isolated phenes in a sufficiently varied natural population necessarily either show specific tendencies in their distribution within the limits of the species habitat or change over time (in paleontological material).

Now, using concrete examples, I will consider the order of isolating phenes that has just been described.

Example I. The purpose of the investigation was to analyze intraspecific variability of coloring (color) of the integument of the sand lizard to elucidate the features of the process of microevolution at the level of a separate population, group of populations, and the entire species as a whole. Thus traits from the smallest to the largest scale were of interest.

1. A study of the variants of coloring according to color of integument of several natural populations showed the presence of a wide spectrum of gray-green and yellow-brown tones in the dorsal coloring, white-bluish-green in the abdomen coloring, and white-yellow in the throat coloring.

2. A comparison of the coloring of young and adult animals showed significant changes of coloring with age: the dorsal coloring of the young was uniformly brown-gray. The coloring of the abdomen and throat changed with age as well. For this reason only sexually mature individuals were used for further analysis; these were easily isolated according to body size.

A study of the coloring of males and females showed a complex picture. In some populations sexual dimorphism in dorsal coloring was clearly expressed (the males were green, the females, brown); in other populations brown and green individuals appeared among both males and females. The coloring of the throat was also complex. As a rule, the males had a more brightly colored throat (bright yellow or even green), but sometimes females with yellow throats were also found. The coloring of the abdomen was greener in males than in females, but it varied in both sexes quite significantly.

3. The very wide spectrum of color variations initially included eight color shades for the dorsal coloring (dark green, bright green, light green, yellow, light brown, reddish, brown, dark brown)

and two or three shades of abdominal and throat coloring. To confirm the correctness of classifying an individual in a given coloring group, special color charts were constructed. But experience with the charts demonstrated a subjectivity of perception in different observers for the initially selected range. All investigators agreed unconditionally only on the selection of green and brown in the dorsal coloring (later, as very rare variants, red—brick-colored—and black animals were found). Green and brown tones in dorsal coloring were selected as discrete traits, and specific consideration of the abdominal and throat coloring was given up completely.

4. The data in the literature on the study of hereditary coloring in the sand lizard pertained to the Mendelian character of inheritance of brick-red dorsal coloring. Data on genetic coloring of lizards of another genus (the American *Sceloporus graciosus*) showed a high hereditary dependence of sandy-red dorsal coloring.

5. Indirect data on the inheritance of coloring under natural and experimental conditions are practically nonexistent because of the significant polymorphism in coloring and the noncomparability of coloring in male and sexually immature individuals. Single black individuals were observed in both young and adult animals in the same habitat (this was a basis for assuming a close relationship).

6. Analysis of frequency of occurrence of green and brown dorsal coloring in males and females exhibited a certain regularity on the whole: in the north, both males and females were brown; in the south, practically all males were green, while both brown and green individuals were found among the females. In other words, a regular change in frequency of occurrence of green and brown coloring was established for both males and females from south to north, while the character of this pattern differed in males and females. Thus, the coloring phenes showed a difference on the scale of large sections of the species habitat. Analysis within separate populations showed no specific results for these traits—their distribution was chaotic.

Figure 4.7 shows certain phenes of the sand lizard, isolated according to pattern of dorsal coloring.

Example II. This example pertains to an analysis of the variability of sutures in the skull of the harp seal. The problem was to

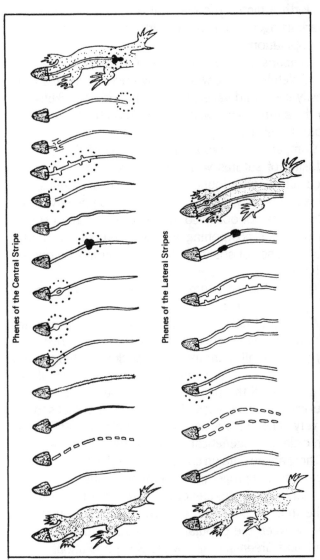

Figure 4.7. Phenes of the central stripe and lateral stripes of the sand lizard *Lacerta agilis*.

Phenes of central stripe: no stripe; none; unbroken stripe; broken stripe; dark stripe; blurred stripe; stripe forming a fork at head; stripe thickening at head; stripe forming a loop at head; stripe broken by spot; zigzag stripe; stripe not reaching head; stripe with processes; stripe fusing with lateral stripes; stripe terminating abruptly in caudal region; stripe interrupted by spot.

Phenes of lateral stripes: no lateral stripe, none; unbroken stripes; broken stripes; stripes continuing on head; zigzag stripes; stripes with processes; stripes interrupted by spots; stripes forming loops. (Drawing by A. Bazan) find evidence of populational independence of three groups of this species in the North Atlantic. The material collected included skulls of animals of various ages (from two to thirty-six years), caught in all three probable populations.

1. Comparison of all observed variability showed that sutures were completely closed, without any traces of joining of the bones, were easily seen, and varied in form. In addition, skulls were found in which the sutures were well developed on the right half, but were absent from the left half (figure 4.8).

2. Comparison of skulls of individuals of different ages showed that closing of sutures was not directly connected to age: clearly expressed sutures were found among the oldest animals, while in some cases sutures in young animals were completely closed. Sexual dimorphism was not observed in the development of sutures. All this was a basis for postponing for later analysis the initially isolated traits of presence or absence of sutures, sutures on one side only, and their form.

3. The presence of sutures was a discrete trait: no animals were found with partial sutures.

4. Analysis of data on genetics showed that some lines of house mice differ in degree of closure of sutures when they reach sexual maturity (in all small mammals sutures close with age). From anthropology, we know of cases of hereditary preservation of non-closed sutures of the skull throughout life. We have data in the literature on the existence of nonclosed sutures in the skulls of old individuals of very different species of mammals, including many carnivores of an order phylogenetically close to the pinnipeda.

5. There are no data on comparing the character of sutures of the skull in parents and progeny in natural populations.

6. Analysis of the frequency of occurrence of various types of sutures in the three populations of harp seal showed that there were differences between the population of the Newfoundland seal, on the one hand, and Greenland and White Sea populations, on the other. These differences were disclosed in a comparison of the occurrence of nonclosed sutures, and not of the forms of the sutures. The phene of presence of sutures was a marker that made it possible to confirm the independence of one of the populations studied.

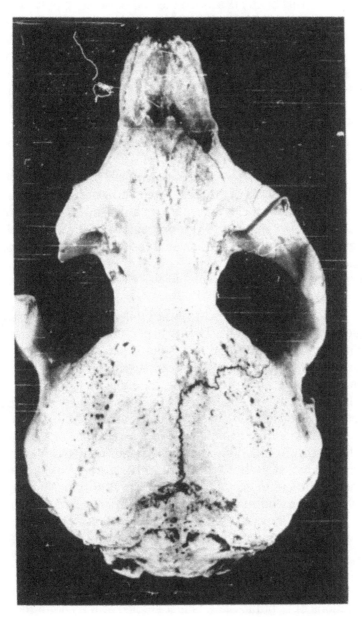

Figure 4.8. Skull on an adult male harp seal *Pagophilus groenlandicus*, with a suture developed only in the right half of the skull. (Photograph by A. V. Valetskii)

This briefly described method of isolating phenes in zoological material can also be applied to work with plants. In isolating phenes, it is important to make sure of having an adequate population.

To resolve any comparatively simple problem in population studies, sometimes only a single phene is needed—for example, in cases where a phene is present in a high concentration in one population, but absent in another, or present in insignificant concentrations. In these cases, we already have a basis for forming a hypothesis on the existence of boundaries between the aggregations of individuals being studied.

But Vavilov's work with various plant species, our work with lizards, and an analysis of the most successful cases of using phenetic methods on other species show that for a dependable and complete characterization of the whole range of spatial variability (from isolation of separate families to isolation of subspecies), it is necessary to consider several dozen, or even several hundreds, traits. Then we will have full assurance that any problem of population analysis will be successfully resolved.

There is no need to worry that time will be wasted on isolating phenes. All that has been said pertains specifically to studies of some group that has not been studied earlier for the purpose of isolating traits which could be accepted as phenes. In the groups where such studies have been made, the situation is quite the opposite. If phenetic material has already been collected for some species (by other investigators, in other places, and at other times), then every subsequent work may potentially lead to ever more interesting and significant conclusions. This is completely analogous with laying out a picture in mosaics: after adding one or two elements, what formerly seemed to be pieces scattered in disorder fall together into a total, sensible picture, and complex designs and patterns become apparent.

Let me cite just one example pertaining to the study of the phene "left turning of the shell" in land snails, *Bradibaena* (*Eulota*) *lantzi*, in the foothills of Alatau beyond the Ili River (Kazakhstan). Left turning is the typical phene and has a monogenic base in all other species of molluscs studied. The distribution of this phene in the populations of Central Asia was first studied at the end of the 1930s and again at the end of the 1960s. Data showed that in one group of the population, the territory occupied by individuals with this phene decreased by many times, and in the other group, it shifted by several dozen kilometers (figure 4.9). Every new study of this interesting natural experiment will be progressively simpler and more informative, and will make it possible to elucidate the dynamics of the genetic composition of the populations.

Scale of Phenes

Phenes may occur on various scales. Some phenes mark single individuals, the smallest of any possible natural groups, the progeny of one pair of parents. Such "family" phenes were the protruding lips of the Hapsburgs and the abnormal finger on the hand of the dukes of Shrewsbury.

Several thousand individual sand lizards were surveyed in the comparatively small area of several square kilometers. All the features of coloring (pattern) were noted, as well as the form and location of scutella on the head and other parts of the body. Several

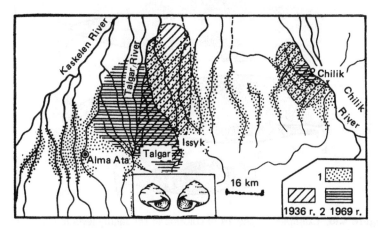

Figure 4.9. Distribution of right- and left-turning forms of the land snail *Bradibaena* (*Eulota*) *lantzi*, and change in habitat area of left-turning forms in the Alatau beyond the Ili River over the thirty-year period 1936–1969. (Yablokov and Valetskii 1971)

dozen traits were considered in field conditions among which, as we assumed, phenes could be present. Among a significant number of individuals examined, only three had a superanal scutellum in the shape of a pentahedron of a unique form. All three of these individuals were caught within a space of only several meters from each other. The only possible explanation of this fact is that the superanal scutellum was a trait on the family scale.

Zoologists learned to recognize by sight individual weasels according to the combination and form of separate spots on the head, throat, and abdomen. Recently we learned that it is possible even at a distance to recognize lions under natural conditions by the distribution of mustache bristles.

A remarkable example of family phenes has been described for sperm whales. Among more than 5000 individuals examined, 2 females were found with a wholly unique pattern never observed before or since: a pronounced perpendicular white stripe on the side of the body (figure 4.10). These females were caught together from one group, and one of the females carried an embryo with exactly the same coloring. This remarkable discovery made it possible to

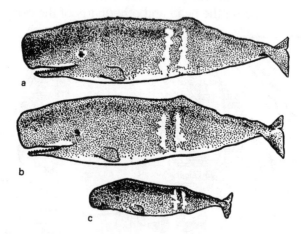

Figure 4.10. Examples of phenes of coloring at the family level in sperm whales *Physeter macrocephalus*, caught from one group (harem): a and b. two adult females; c. embryo. (Drawing by G. Veinger)

resolve a sperm whale biology problem that had up to that time been the subject of heated arguments among specialists. It demonstrated that females belonging to a harem—a group of 10–30 females with 2 or 3 large males—may be related, and the whole harem may be a family group. Finding mothers and offspring with the same rare pattern, indicating the hereditary nature of this trait, was a basis for regarding it as a true phene. Thus, the smallest specific rare variations of any kind of trait present the investigator with phenes at the family level.

In any adequately broad study, it is possible to observe phenes of an even larger scale. In animals these phenes mark demes, groups composed of several family groups (or even larger and more complex groups), that occupy a specific space within a population and preserve relative independence over several generations. Such groups were found in the sand lizard in the Altai Mountains: several of the phenes considered varied in concentration in different valleys between neighboring hills. In other cases, investigators found statistically reliable differences in the distribution of various phenes for groups of individuals within populations of a number of mouselike rodents and other species of animals.

In the following chapter I will analyze examples of isolating separate populations and groups of populations according to the frequency of distribution of various phenes. Extreme cases of this type are phenes on the scale of species or the scale of even larger groups.

It should be noted that rare phenes make it possible to distinguish some very small groups of individuals within a population, as well as quite large groups. An example of this is the phene of perfectly black body coloring (complete melanism) found in the sand lizard on the Kalbin ridge of the Altai Mountains and in the Barnaul area. This phene allows us simultaneously to distinguish the deme and the whole large group of the Altai populations.

Obviously, phenes found in average or high concentrations are suitable only for comparing quite large groups of individuals.

Thus, phenes occur on different scales, from individual and family to species. They may unite several species as well; in paleontological material phenes may characterize separate small branches of the tree of life.

Mammalian Phenes in Greater Detail

What has been said above pertains to general principles of isolating phenes. Now we will look only at one single group of animals from the phenetic point of view: mammals. Undoubtedly, in the not-too-distant future phene atlases or catalogs for various groups of plants and animals will be indispensable equipment in many laboratories. At present no such atlases exist, and the enumeration presented here of probable, or already discovered, phenes in various systems of mammalian organs will be a basis for a more complete concept of the potentials of the phenetic approach.

Groups of traits are listed below, the hereditary nature of which has been confirmed either experimentally in genetic research or indirectly by data on their marking various groups of individuals within a species.

Form and size of body. The principal discrete variations are dwarfism and gigantism, shortening of extremities (for example, the famous Ancon sheep cited in Darwin's *Origin of Species*), "turned nose," severe shortening or elongation of the snout and/or head, short tail and taillessness, loss of a finger, merging of digits on the front or hind feet, or additional digits. According to the form of the body we can now isolate several dozen phenes.

Coloring. Six principal and many complementary genes affect the color of mammalian integument and determine the appearance of dozens of discrete traits according to color. No less than ten genes are connected with mottled or striped coloring. On the whole, with material that has been thoroughly studied genetically (mouse, rat, rabbit), several hundred discrete color traits that can be regarded as phenes have been identified. Among these are mottling of various colors, shades, forms, and sizes: light on a dark ground; dark on a light ground with sharp boundaries, with blurred boundaries, with a light boundary; dark and light; small and large; round and angular; complex in form (cruciform, heart-shaped, lyre-shaped, stars, rings, rosettes, etc.). Head coloring includes black; white; light above, dark below ("badger type"); dark above, light below (there is a breed of sheep in which the combination of the last two types of coloring in crossbreeding results in a black head); coloring of the mask, e.g.

"eyeglasses." Ears are dark, light, or dark or light at the tips. Other characteristics of the head include dark crown, dark or white spots on the cheeks; a white tuft of hair on the head; dark spots near the ears, a "star" on the head; graying of the hair with maturity, etc. Even in the color of cetaceans, dozens of phenes can be isolated (figure 4.11).

Structure of hairy integument. Some of the discrete traits

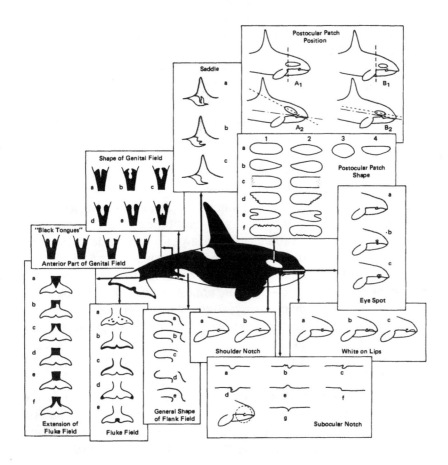

Figure 4.11. Components of the *Orcinus orca* color pattern with some observed shape and positional variants. (Evans, Yablokov, and Bowles 1982)

observed in the hairy integument of the black rat are locks of hair
on the spine, curls of hair on the back and thighs, absence of hair on
certain parts of the body, very long hair, tufts of hair on the ears, long
hair in tufts over the whole body, "plush" without obvious hairs, wavi-
ness, and twisted or bent bristles. Probably about a hundred phenes
could be isolated in various mammals according to features of the
structure of the hairy integument.

   Derivatives of the integument. Dozens of variants in each species studied with respect to the location and occurence of bristles, are
recognized. In large predators, the location of bristles is one of the
convenient traits for identifying individuals.

   In mammals we recognize many discrete variants in the
location and number of nipples on the mammary glands, scaliness of
the tail in rodents, and papillary patterns in man and primates; variants
also appear in the location of plantar protuberances and in papillary
patterns of the nasal speculum in hoofed animals, and of flippers in
seals.

   On the whole, probably 500–700 phenes can be identified
among the features of the integument.

   Skeleton. Quite a number of phenes can be identified
among the features of the skeleton: a multitude of small variations of
the skull (hundreds of variants in quite large skulls); the dental system,
which we are arbitrarily considering in this chapter (specifically, the
dental system of man, which is the best studied thus far, although it
is not the most complex—there are approximately 400 variants, many
of which are obviously phenes; features of tooth structure in rodents
make it possible to isolate dozens of phenes); the fore and hind regions
of the extremities; and the axial skeleton. No fewer than several
hundred hereditary variations have already been isolated in all parts
of the skeleton.

   Digestive system. Dozens of variants can be found in the
form, number, and location of various types of gustatory papillae on
the tongue. Discrete variability appears in the form of the liver lobes,
pancreas, gall bladder, and bile ducts (even an absence of the gall
bladder, a trait that is common to intraspecific variability in many
species). The boundary between the horny and glandular epithelium
of the upper and lower intestine, the form of folding of the glandular

epithelium in different parts of the intestines, etc., constitute other traits. The digestive system is comparatively little studied from the phenetic point of view, but 100–150 phenes can probably be isolated in it.

Musculature. Intraspecific variability of the muscle system has hardly been studied, but, without exception, all studies of sufficiently large samples show a remarkable variability in the structure and topography of even the most common and well-developed muscles. In the future dozens of phenes will probably be isolated.

Respiratory system. Lobulation of lungs (on the basis of a population of seals, it was demonstrated that these phenes are useful on the interpopulation level), branching of the bronchial tree and the lung arterial tree, and the structure of diverticula of the larynx and nasal passages indicate the presence of discrete variations. In the future many more phenes will probably be isolated.

Urogenital system. Discrete variations occur in the topography of the kidneys and suprarenals (in the case of lobed kidneys features of joining of the separate lobes), mouths of the ureters, pattern of the epithelium that lines the urinary bladder, forms and location of the os penis and os clitoridis, folds of the vagina, and forms of the testes and ovaries. Dozens of phenes can also be isolated in this system.

Nervous system and sensory organs. We recognize discrete variants in coloring and anomalies of structure of the eyes (no fewer than 20–25 different phenes) and in the structure of the outer ear (tens of phenes in man alone); variants exist in the topography of separate nerve trunks and networks (without exception, all the series studied show great variability). From the phenetic point of view, the nervous system has been little studied, and thus far, several dozen phenes can be isolated.

Circulatory system. A great number of hereditary variants in the topography of large and small blood vessels—both arteries and veins—in various parts of the body is known; the form and topography of the spleen vary. The circulatory system, together with the organs of the integument and skeletal systems, is one of the most promising from the position of phenetic research and contains potentially hundreds of comparatively easily recognized phenes.

Karyotype. A great number of variants are recognized for the structure of chromosome sets: intraspecific chromosome polymorphism is evidently characteristic for all species of mammals without exception, as many studies have demonstrated.

In mammals on the whole, no less than two or three thousand hereditarily determined, morphological, discrete traits, or phenes, are now known. Undoubtedly, with the development of the phenetic approach, the number of these traits will grow rapidly and reach thousands for each organ system.

The possibility of such a sharp increase in the number of documented phenetic traits is indicated by the status of research on biochemical properties—called "biochemical phenotypes," the phrase applied with great precision more and more often to the results obtained in biochemical studies with electrophoretic, immunological, and other methods. Thirty or forty years ago, single biochemical traits subject to intraspecific variability were known; today we know many hundreds of such traits in many species.

A great number of phenes should be discovered in population-physiological studies. For example, as a result of controlled selection, lines of rats were developed with high and low blood pressure; these differences are connected with variations in one gene that controls the synthesis of corticosterone. Discrete genetic variations in growth hormone content, gonadotropines, insulin, vasopressin, catecholamine, prostaglandin, and many other hormones are known in a whole series of mammalian species. Now such studies are being conducted more extensively, and in the field of physiology, we can apparently expect several dozen or perhaps even hundreds of phenes to be recognized in the next few years.

Significantly, more phenes are now being isolated in the behavior of mammals. The development of ethology uncovered extremely small hereditary variants in behavior that would have been difficult to imagine a hundred years ago. (Moreover, we all know well the innate small peculiarities in man passed from generation to generation—distinctive walks, kinds of gestures, etc.). Greater amounts of data are also being accumulated on the voice phenes of mammals.

Thus, if we add the biochemical, physiological, and ethological phenes to the groups of morphological phenes mentioned above, today there are thousands of precise traits at the disposal of investigators. The selection of several dozen of these for specific population research is now possible for practically any mammalian species.

Are Phenes Adaptive?

In one form or another this question is almost always brought up in a discussion of works in the field of phenetics. I spoke above of the phene of left and right turning in a species of land snail. It turned out that individuals carrying this phene differ in the way they use energy reserves. This indicates that left turning is not a neutral trait but is under the control of natural selection. Recently Sverdlovsk zoologists demonstrated that differences between nonstriped and striped forms in other species of this genus correspond closely to differences in intensity of CO_2 formation in respiration (Khokhutkin and Dobrinskii 1973).

Many works of this kind can be cited; they all lead to the general conclusion that in all specially studied cases, individuals with different phenes always exhibit certain adaptive differences. The red coloring phene in some populations of two-spotted ladybird beetles, *Adalia bipunctata*, is somehow connected with the capability to tolerate cold better during hibernation, and the black coloring phene, with more intensive breeding in the summer. The yellow color phene in the California field mouse is linked with better survivability at a low temperature and with better breeding on a diet of green food (and not seeds) than that of individuals with typical gray coloring. The gray coloring phene in the American cricket frog is connected with the organism's resistance to loss of moisture and its better ability to resist certain infections.

The principle of pleiotropy (*polypheny*), the effect of one gene on many traits, implies that among individuals carrying different phenes, there must be many differences in biological traits and properties. Some phenes probably also have direct adaptive significance. In other cases the phenes have indirect adaptive significance connected with important biological traits and properties in the development of the organism.

An indirect, but quite definite, confirmation of the adaptive character of phenes is the data on the combination of different discrete variants in the structure of the grinding surface of the teeth in bank voles. Five variants in the structure of the third upper molars and four variants in the structure of the first lower molars in this species, isolated by Saratov zoologists, could theoretically yield 400 combinations. But in most of the individuals studied there was not more than 10 percent of this number. If the isolated structural variants made no difference to the organism, then we might expect the most various combinations. But this was not so, and evidently most of the combinations were unfavorable for the organism.

Another important index of adaptive "nonindifference" in phenes is the remarkable stability of their frequency in natural populations. A comparison of the frequency of certain phenes of dorsal patterns in the sand lizard in collections made in one and the same place, but at intervals of seventy years, demonstrated its stability. Other examples abound of the preservation of stable phene frequency (differences between lines) over dozens of generations in many species of drosophila, laboratory mice, rabbits, rats, and practically all laboratory animals and other animals bred in captivity; plant cultures and domestic animals show similar patterns. If the phenes characterizing all these genetically more or less uniform aggregates of individuals did not have adaptive significance, then it would be difficult to imagine their being maintained in a stable state even over several generations.

It is important to emphasize that ignorance of the concrete adaptive significance of a given phene must not serve as an obstacle to its use as a trait, a marker of natural groups of individuals. This conclusion is methodologically important since it prevents us from studying concrete adaptations that simply can never be fully elucidated. This conclusion seems somewhat unexpected perhaps, but it follows inevitably from the limitations of our knowledge at any given moment. I offer only one example here.

For a long time it was thought that the wings of bats are only an organ of flight. But a more careful observation under natural conditions showed the important role of bat wings as a net for catching prey. Soon the important role of their wings in thermoregulation was disclosed: they are the only large parts of the body not covered by fur, and they are cooled significantly in flight. On the other hand, the animal may fold its wings during rest and in this way form a kind of additional leathery thermoisolating integument. Thus, besides making flight possible, the wings of the bat also function as a net and as a thermoregulator. In addition, physiologists suggest that the skin of mammals is an organ in which vitamin D synthesis takes place, and this process occurs only under the influence of sunlight. Perhaps in bats, the large wings, with their enormous surface of skin, compensate to a degree for an inadequate production of vitamin D since they are not usually exposed to sunlight, these animals being either twilight or nocturnal animals. In the future, other functions will certainly be discovered for their wings (Yablokov 1974). And this process of deepening knowledge will proceed endlessly, paralleling the increase in knowledge of the structure and functioning of everything that is alive.

The situation described in the bat example illustrates the principle of multifunctionality of any organ or structure, the well-known existence of a multitude of functions in every organ or body part in the organism of any being. But since our knowledge will never exhaust this multitude, does not the requirement of preliminarily explaining the adaptive significance of one phene or another and only then using such a trait in research seem mistaken or, more precisely, unnecessary?

To study natural aggregations of individuals using discrete traits-phenes, we need only know in general that the given trait of the organism, and this includes phenes, must have some direct or indirect adaptive significance.

REFERENCES
Auerbach, C. 1961. *The Science of Genetics*. New York: Harper & Row.
Berry, A. C and R. J. Berry. 1971. Epigenetic polymorphism in the primate skeleton. In B. Chiarelle, ed., *Comparative Genetics in Monkeys, Apes, and Man*, pp. 13–41. London and New York: Academic Press.
Berry, R. J. 1968. The biology of nonmetrical variation in mice and man. In D. R.

Brothwell, ed., *The Skeletal Biology of Early Human Populations*, pp. 103–133. London: Pergamon.

Evans, V., A. V. Yablokov, and A. E. Bowles. 1982. Geographic variation in color pattern of killer whales (*Orcinus orca*). *Rep. Int. Whal. Comm.* 32:687–694.

Haecker, V. 1925. Aufgaben und Ergebnisse der Phaenosgenetik. *Bibl. Genet.* 1:93.

Khokhutkin, I. M. and L. N. Dobrinskii. 1973. Differences in gas metabolism in two morphs of land snails *Bradibaena schrencki* (Midd.) (In Russian.) *Ekologiya* 6: 90–93.

Plate, L. 1910. Vererbungslehre und Deszendenztheorie. *Festschr. f. R.Hertwig*, II. Jena:Fisher.

Shire, J. G. M. 1976. The form, uses, and significance of genetic variation in endocrine systems. *Biol. Rev.* 51:105–141.

Sjovold, T. 1977. Nonmetrical divergence between skeletal populations. *Ossa*, 4, suppl. 1:1–133.

Timofeeff-Ressovsky, N. V. 1931. Gerichteter Variieren in der phanotypischen Manifestierung einiger Genmutationen von *Drosophila funebris*. *Naturwissenschaft* 19:493–498.

Timofeeff-Ressovsky, N. V. and S. R. Zarapkin. 1932. Zur Analyse der Formvariationen. *Biol. Zbl.* 52(3):138–147.

Yablokov, A. V. 1974. *Variability of Mammals*. (Translated from Russian.) Revised ed. New Delhi: Amerind.

Yablokov, A. V. and A. V. Valetskii. 1971. Changes in habitat area of left-turning forms of *Eulota lantzi* Lind. in the Alatau beyond the Ili River in the last decade. (In Russian; English summary.) *Zool. Zh.* 49(1):121–130.

FOR ADDITIONAL READING

Berry, R. J. 1969. Nonmetrical skull variation in two Scottish colonies of the grey seal. *J. Zool.* (London), 176:11–18.

Corruccini, R. S. 1978. An examination of the meaning of cranial discrete traits for human skeletal biological studies. *Amer. J. Phys. Anthrop.* 40:425–446.

Dempster, E. R. and I. M. Lerner. 1950. Heritability of threshold characters.*Genetics* 35:212–236.

Drummond, D. C. 1970. Variation in rodent populations in response to control measures. In R. J. Berry and H. N. Southern, eds., *Variation in Mammalian Populations*, pp. 351–378. London: Academic Press.

Dunn, L. C., A. B. Veasley, and N. Tinker. 1960. Polymorphisms in populations of wild house mice. *J. Mamm.* 41(2):220–229.

Eremina, I. V. 1978. Methods of isolating phenes of the grinding surface of molars in field mice. In *Physiological and Populational Ecology of Animals* 5(7):60–68. (In Russian.) Saratov: Saratov University Press.

Gruneberg, H. 1952 Genetical studies on the skeleton of the mouse, IV: Quasicontinuous variations. *J. Genet.* 51:95–114.

Gruneberg, H. 1952. The genetics of the mouse. *Bibl. Genet.* 15–650.

Hilborn, R. 1974. Inheritance of skeletal polymorphism in *Microtus californicus*. *Heredity* 33(1):87–121.

Howe, W. L. and P. A. Parsons. 1967. Genotype and environment in the determination of minor skeletal variants and body weight in mice. *J. Embryol. Exper. Morph.* 17(2):283–292.

Larina, N. I. and I. V. Eremina. 1982. Certain aspects of the study of pheno- and genotype of species and interspecific groups. In A. V. Yablokov, ed., *Phenetics of Populations*, pp. 56–68. (In Russian.) Moscow: Izd-vo Nauka.

Nevo, E. 1973. Adaptive color polymorphism in cricket frogs. *Evolution* 27(3):353–367.

Puček, Z. 1962. The occurrence of Wormian bones (Ossicula wormiana) in some mammals. *Acta Theriol.* 6(3):33–51.

Robertson, A. and I. M. Lerner. 1949. The heritability of all-or-none traits: Variability of poultry. *Genetics* 34:395–411.

Robinson, R. 1970. Homologous mutants in mammalian coat color variation. In R. J. Berry and H. N. Southern, eds., *Variation in Mammalian Populations*, pp. 251–270. London: Academic Press.

Thompson, J. N., Jr. 1977. Analysis of gene number and development in polygenic systems. *Stadler Symp.* 9:63–82. University of Missouri, Columbia.

Vigorov, Y. L. 1979. Possibilities of analyzing nonmetrical variability of various bones of the skull. In *Quantitative Methods in Ecology*, pp. 94–100. (In Russian.) Sverdlovsk.

Yablokov, A. V., A. S. Baranov, and A. S. Rozanov. 1980. Population structure, geographic variation, and microphylogenesis of the sand lizard (*Lacerta agilis*). In M. K. Hecht, W. C. Steere, B. Wallace, eds., *Evolutionary Biology* 12:91–127. New York: Plenum.

CHAPTER FIVE

Methods of Studying the Phene Pool

In the 1920s in the Moscow school of genetics, headed by N. K. Kol'tsov, and in the Leningrad school, headed by Yu. A. Filipchenko, population-genetic problems were studied intensively. As Timofeeff-Ressovsky recalls, the term "phene pool" appeared for the first time in these groups of geneticists. It is possible that Serebrovskii, who introduced the concept of a gene pool in 1927, played a decisive role in this.

The phene pool is the totality of phenes of any assembly of individuals. The number of alleles is a definite number; it depends on the number of genes, the number of traits being practically infinite (it is always possible to increase the number—this depends on the desire and persistence of the investigator). For this reason every study of the phene pool will only approximately establish the allele pool as a totality of all alleles.

In this chapter I will consider methods of expressing the phene pool and two principal directions for investigating this pool: studying the spatial structure and dynamics of the genetic composition of a population. In other words, I will be speaking of the study of the elementary evolutionary structure and the effect of various elementary evolutionary factors. In conclusion, I will discuss the various methods of describing a phene pool.

Studying the Structure of a Population

I have already stressed more than once the importance of a precise characterization of the elementary evolutionary unit, the population. One of the most important population characteristics is the spatial structure of the population. The distribution of genetic information within the limits of the whole population, the rate of evolutionary change, and all other microevolutionary developments depend to a significant degree on what kinds of groups of individuals the whole population is divided into and what the relationships are between these groups.

Before the development of the phenetic approach, the structure of a population was determined basically according to eco-logical observations, the character of the distribution of individuals in space. (This is determined either by tagging and capturing the tagged animals, by long-term observation of specific animals, or by quite indirect methods such as the analysis of the stomach content of animals caught in various biotopes.) By using marker traits of geno-typic composition, phenetics makes it possible as a rule to elucidate the structure of a population with immeasurably greater precision and significantly less effort. Let us consider several typical examples.

A study of the flipper of the white dolphin (*Delphinapterus leucas*) made it possible in the early 1960s, for this author to describe several dozen variants in the location of carpal ossicles. Some variants were very similar, others, very different. Examining the sketches in my field journal, I found, quite by chance, that in white dolphins caught in one herd, the structure of the ossicles had many more similar characteristics than in individuals from different herds. Special anal-ysis confirmed this observation: small herds of white dolphins, in-cluding basically females and young dolphins of various ages with single males, were groups that were surprisingly similar with respect to this trait. From a phenetic point of view, the carpal structure of the dolphin is no less precise a trait than, for example, the hereditary trait of fingers that are grown together. Without doubt all the basic variants of carpal structure are strictly genetically determined. This means that a group of white dolphins with similar flipper structure represents something in the nature of a family group. Only in this way can we understand the observed phenetic similarity. A study of the flipper of the white dolphin demonstrated the family character of the organi-zation of the dolphin herd (Bel'kovich and Yablokov 1965). A large family such as this includes all the daughters of one old female and their children. When males become sexually mature, they leave the group and join separate male shoals, keeping apart for most of the year from the family herds. The males join such a family only during the breeding period. It is led, evidently, by an old female so that there is a true matriarchy in this species of dolphin.

While isolated morphological types of flipper structure cannot, of course, be labeled phenes (these are complex traits), the phenetic approach here was fully apparent and made it possible to reach conclusions which would have been impossible with other methods (perhaps a long-term tagging and observation of these animals over decades would have given the same result).

Another example was described in 1970 by the American biologist R. K. Selander, and involved the ordinary house mouse. On an old poultry farm in Texas, in a large barn measuring 20 × 60 m heaped up with bird droppings, lived 3000 mice. Hereditary variants of four different proteins (enzymes) in the blood of each of 826 trapped individuals were determined by electrophoresis. Approximately thirty enzymes were studied, but most of them showed no differences within the population of this barn (although in comparisons with mouse populations of different regions of the United States many of them exhibited regional geographic variability).

As a result of the studies, it became clear that in the one barn where, it would seem, the animals must interbreed freely, the whole mouse population was divided absolutely into a number of phenetic groupings. The isolated phenetic groups indicated the genetic variability of the mouse population of the whole barn and the genetic relationship of mice within the isolated small groups. These small homogeneous groups had similar genetic characteristics, and within them a high degree of panmixis was found. Each of these groups contained 6 to 80 individual mice.

The work of a Soviet investigator, N. I. Larina, demonstrated that according to concentration and complements of phenes of the skull, it was possible to distinguish the populations of field mice inhabiting two haystacks standing side by side in a field. Evidently these were also groupings composed of a few related families living together.

Quite a number of such studies have been done. Groups of bank voles, living in a dense forest at a distance of several dozen meters from one another, can be distinguished according to the concentration of phenes of the skull (figure 5.1). Groups of mice living in different parts of large buildings can be distinguished according to the same kind of trait, as can groups of black rats from neighboring houses in some Indian cities; these differ markedly in frequency of mottled individuals, etc. (figure 5.2).

In all the mammals studied, the population is made up of comparatively small groups of individuals, and the groups are made up of a small number of related families living together, united by a genetic relationship. The most various terms have been proposed for such groups in different species, but the term "deme" is used most frequently.

From these examples, it is apparent that phenetics helps to disclose the genetic structure of a population and indicates the groups into which the whole population is divided.

In some cases, the phenetic study makes it possible to descend even lower along the hierarchic ladder of intrapopulation associations and detect characteristics of the brother-sister or parent-child type. One such study was done by Larina, Golikova, and Eremina (1976). Figure 5.3 shows the situation at the time of capture in a trap of ten yellow-throated mice living in a section of a mountain beech forest in the vicinity of Dilizhan, Armenia, and the phenetic formulas (phenocomplexes) of the dental system. Among these animals, it was possible to distinguish with a great degree of reliability, according to the character of individual phenocomplexes, a group of seven animals that were a family of parents and children.

A clear example of illuminating the genetic relations of close relatives in natural populations is given in the preceding chapter. It pertains to color phenes of sperm whales that made it possible to isolate two sisters in one harem. (see figure 4.11). In other chapters of this book, the reader will find a number of examples of such a population structure arrived at by isolating and describing separate phenes and phenocomplexes.

The possibilities for individual phenetic recognition of animals is much broader than one might think. In Slimbridge, England, at the Wildfowl Trust directed by Sir Peter Scott, more than a thousand small swans (*Cygnus berwichii*) gathering here for the winter can be

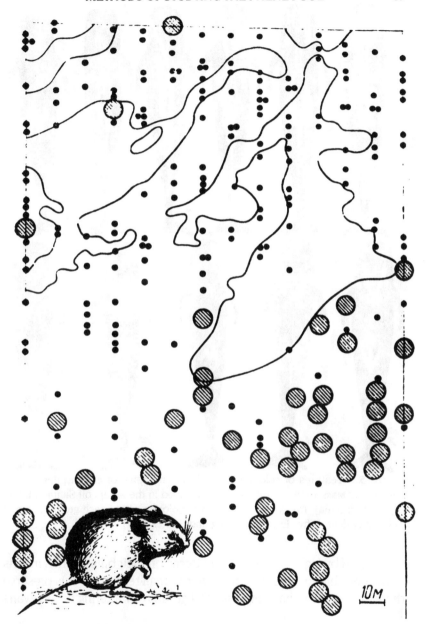

Figure 5.1. Locations of trapped bank voles, *Clethrionomys glareolus* (dots), and locations of occurrence of individuals with the "interrupted dental row" (circles) on
a 9-hectare area in a mixed forest on the Valdai upland (Upper Volga region). Fine
line represents contour of the marshy forest. (After Turutina 1982)

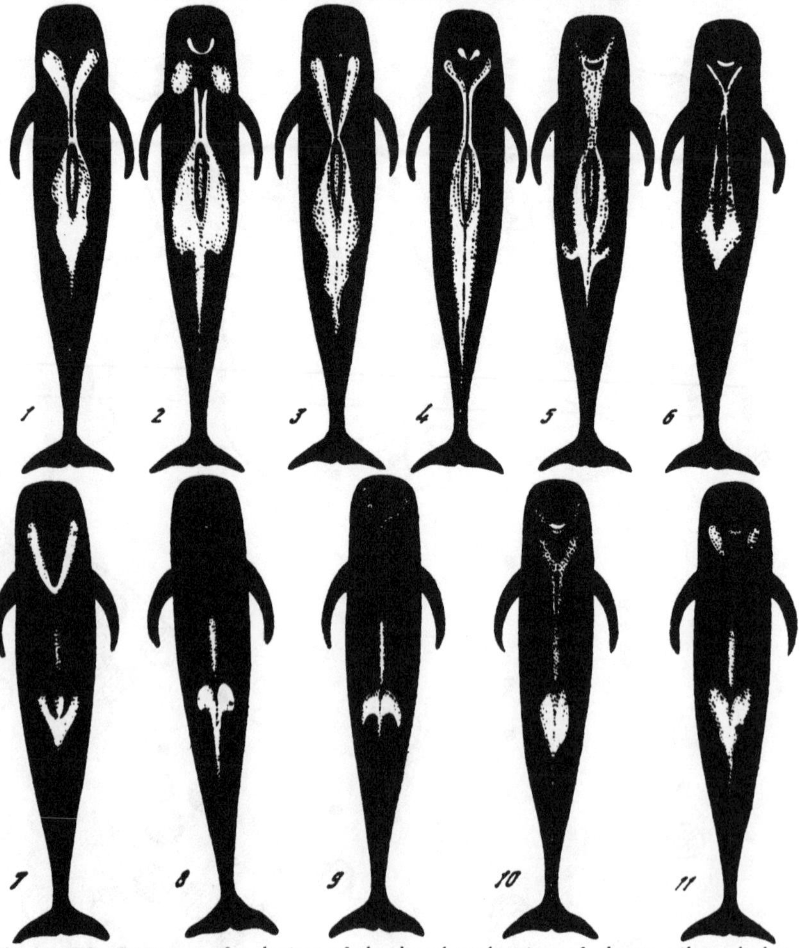

Figure 5.2. Features of coloring of the head and spine of eleven pilot whales, *Globicephalus melas*, in two natural groups observed in the waters off Santa Catalina Island (California). The general similarity in coloring within each group (1–6 and 7–11) is evident. (After Evans, Yablokov, and Bowles 1982).

recognized by sight according to patterns on their bills! By this means it was possible to determine that on the small territory of this preserve there are no fewer than seven related groups of wintering birds (Evans 1977).

Studying larger associations of individuals than demes, phenetics makes it possible to detect genetically interrelated groups

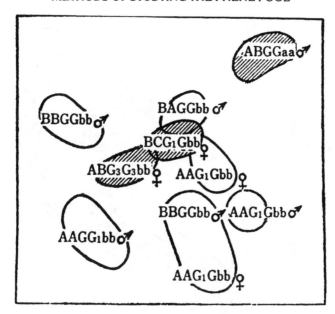

Figure 5.3. Individual sites of ten yellow-throated mice, *Apodemus flavicollis*, with different phenetic formulas in a mountain beech forest in the vicinity of Dilizhan, Armenia. (Larina, Golikova, and Eremina 1976)

of demes and can even isolate populations (figure 5.4, table 5.1). In the N. K. Kol'tsov Institute of Developmental Biology, data have been obtained on the phenetics of the sand lizard, which represent a hitherto rare example of sequential phenetic research carried out from the very beginning and directed toward the study of typical microevolutionary situations under natural conditions. In one of the populations in western Altai, we examined forty color phenes of several thousand lizards, with a precise linking of each animal to the specific place of capture. We found phenes that characterized two or three neighboring burrows—evidently associations of the brother-sister type. We also located phenes that characterized separate small hollows with a population of several dozen or several hundred individuals (evidently demes) and phenes that could be used to isolate three or four such demes (figure 5.5).

All these examples indicate the kind of population structure—the spatial-genetic (biochorological)—that the phenetic ap-

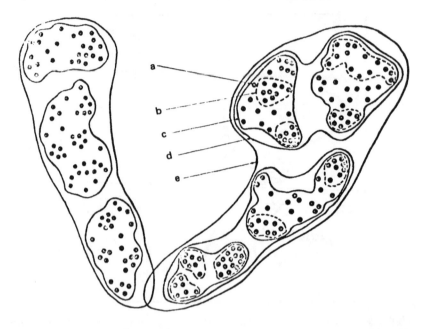

Figure 5.4. Diagram showing the separation of a population into several biochorologic groups: a) family; b) deme; c) group of first-order demes; d) group of second-order demes (population); e) group of populations. See table 6.1 for further details. (Yablokov, Baranov, and Rozanov 1980).

proach can deal with. It is especially important that phentics presents a unique possibility for the study of the genetic structure of natural populations (complementing the spatial structure long and successfully studied by ecologists).

It would be improper, however, to separate genetics and ecology in any way in the study of the structure of a population: phenetic study alone will not give a sufficiently complete concept of it. Thus, for example, ecology shows that phenetically isolated demes of small vertebrates exist for one or two generations, to the season of mass fall reproduction in our latitudes. Then the wave of young migrating individuals fills the whole territory of the population, and new demes are formed that will live to the succeeding population wave. In the study of population structure, phenetics goes hand in hand with ecology.

Table 5.1 Distinctive Features of the Main Levels of Integration of the *Lacerta agilis* Population in the Western Altai

Level of Integration	Basis on which Level was Identified	Number of Individuals	Occupied Area	Level of Gene Exchange with Neighboring Groups per Generation	Survival Time	Proposed Name
1	Rate phenes	Several	0.1 ha	50%	1–2 generations	Family
2	Chorologic data and specific phene frequency	Several dozen	1 to several hectares	Approx. 20%	Several generations	Deme
3	Specific phene frequency	Several hundreds	Several to several dozen hectares	Approx. 3–4%	Tens of generations	Group of demes
4	Chorologic data and specific phene frequency	Several thousands	Several to several dozen hectares	Approx. 0.01%	Hundreds of generations	Combination of groups of demes (population)
5	Chorologic data and specific phene frequency	Tens to hundreds of thousands	Hundreds of thousands of hectares	Undetectably small	Thousands of generations	Group of related populations

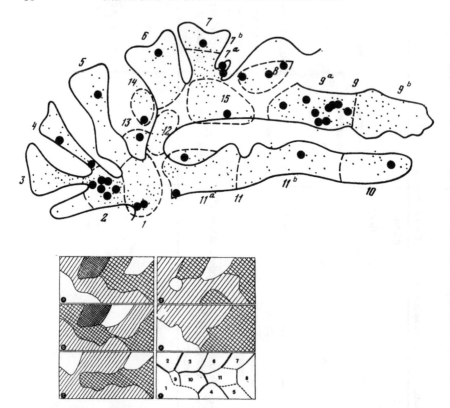

Figure 5.5. An example of a study of the biochorologic structure of a population.

5A. According to the concentration of sand lizards with an interrupted central band (large dots) chorologic groups 1 and 2 can be distinguished with statistical reliability ($p > 0.05$); groups 2 and 3, with $p > 0.01$, and two parts of group 9, with $p > 0.01$. Small dots represent places where lizards were caught.

5B. By superimposition of charts of distribution of zones with the same concentration of various phenes (a–f), it is possible to isolate phenetically homogeneous segments (1–11). (After Turutina 1982; Yablokov, Baranov, and Rozanov 1980).

Nevertheless, we must not forget that the final purpose of such research is the understanding of genetic processes in natural groups of individuals—the elucidation of those currents of genes that unite various gene pools of separate groups into one gene pool, stable over many generations, which are the principal material of microevolutionary changes.

Studying the Dynamics
of the Genetic Composition
of a Population

It was stressed at the beginning of this book that one of the most important tasks of modern population studies is obtaining material on the most various evolutionary situations in natural populations, particularly data on the action and interaction of elementary evolutionary factors. Phenetics may be very useful in disclosing the features and intensity of the action of these factors.

The first way is to clarify the action of natural selection by comparing the concentration of some phenes at different times in the life of a single generation. If an individual is born with some trait which does not change later in the course of ontogenesis (the number and location of scales, venation of wings in insects, skin pattern in primates, etc.), then a comparison can be made of the distribution of this trait in different individuals at different stages of ontogenesis— for example, in adults and the newborn. In a number of cases, a notable shift will be apparent in the frequency of incidence of this trait.

Several years ago, a Soviet zoologist, V. G. Ishchenko (1971), studied the frequency of incidence of various kinds of openings for the passage of blood vessels and nerves in the skulls of polar foxes found on the Yamal Peninsula. Of the ten phenes studied (pertaining to variants of location of the opening in front of the eye), for three, statistically significant changes in frequency of different phenes were found between this year's brood and yearling foxes within a single generation (table 5.2). Similar age changes in a number of other traits

Table 5.2 Changes in Phene Frequency Over a Lifetime of a Single Generation in Polar Foxes (*Alopus lagopus*)

Sex	Age	Phene			n
		C_1	C_2	C_3	
Males	This year's brood	43.1	33.8	23.1	65
	Yearlings	77.8	16.6	5.6	54
Females	This year's brood	53.1	26.5	20.4	49
	Yearlings	90.5	9.5	0.0	42

of the skull were found by the same investigator in ermine *Mustela erminea*.

Natural selection can be discerned in such shifts, but how can they actually occur? Only as a result of eliminating some of the individuals who might have been destroyed by enemies, might have died of disease, and so forth. If the number of those lost was relatively the same for all individuals with various characteristics, no shift in the frequency of traits would be detected. A shift points to the action of natural selection, although, of course, it cannot disclose the specific reason for the death of one set of individuals or another. This example is exactly such an equation with one unknown about which every biologist-investigator must think when only one explanation can be singled out from a multitude of factors and actions.*

The example given above of a change in phene frequency over the life of one generation is not the only possible example. Also of interest is an analysis of differences in frequency of phenes of striata and maculata in a single generation of the frog, *Rana arvalis*. This analysis showed a differential mortality of individuals that were carriers of different phenes. Ecological analysis of this unusual natural circumstance over two years made it possible to determine the direction of natural selection: evolution of the rate of development and decrease in sensitivity to mass destruction (Ishchenko 1978).

Comparing the phenetic characteristics of different phases of ontogenesis within one generation is one way of studying natural selection through phenetic methods.

One way of studying the effect of natural selection is evident when an explanation is required for the specific adaptive significance of one phene or another. Of course, we will never know fully all the adaptations of even one structure, but in a number of

---

* Before leaving this topic, I will add that the most common methodological error in any creative process is that the investigators ponder over "an equation with many unknowns." Such equations do not have unequivocal solutions because attempts to solve them precisely don't make sense. In any investigation, it is necessary to try to pose an equation with one or two unknowns (an equation with two unknowns has a limited number of solutions, and it is sometimes interesting to think about them). The practical conclusion from this discussion is simple: if there are more than two unknowns in our biological constructs and hypotheses, they must be simplified and made more stringent by bringing the number of unknowns down to one or two. Otherwise the whole work will be useless, and it will be impossible to get a definite answer.

cases it is possible to assume confidently that we know the main adaptive significance of an organ or structure.

A study on a North American deermouse species, *Peromyscus*, may be an example of such an analysis of selection. The study of polymorphism of their coats showed that there are brown and pale yellow individuals (Smith and Carmon 1972). A genetic experiment revealed that the yellow coloring phene was a recessive trait and that it should not be present in natural populations in the high concentrations in which it was observed. It was found that the yellow coloring phene is connected with a longer lifespan and more successful reproduction. For this reason, regardless of the fact that the yellow coloring phene is a recessive trait, it is sustained in a high concentration by natural selection. The animals living on a dark land surface are better protected by a dark coat color and the light-colored, yellow forms are more likely to be killed by predators. But the light-colored animals have a longer lifespan; for this reason they leave more progeny, and their concentration is maintained at a high level.

A more graphic example of a phenetic study of the action of selection in nature is the semiexperimental study by the American biologist L. I. Brown of house mice in a grain elevator in Missouri; the results of this study were published in 1965. Brown found a rare mutation, "yellow color + red eyes," in a surprisingly high concentration, up to 30 percent. Genetic studies showed that these traits characterized a homozygous state of one of the recessive genes. Over a period of three months several cats were released into this grain elevator; they caught every one of the yellow mice (such light-colored mice were more noticeable, and the cats were able to catch them more easily). But seven months after the cats were removed, yellow mice were again found in the population. Without a constant elimination of the light-colored individuals by the cats, the recessive homozygotes again emerged among the heterozygous gray individuals. These examples show that the study of even a few phenes makes it possible to reach interesting conclusions regarding the direction of natural selection.

Another way of studying the effect of selection is by comparing the phene pool of various generations. Here we can follow the desired course with few unknowns only if we succeed in isolating traits that are changing in some specific direction over the course of many generations. Actually, since selection is the only directional evolutionary factor, then any directional change in evolution may show the effect of selection.*

The teeth (reticular enamel pattern) of small mouselike rodents are a very convenient material for various phenetic studies. In a series of works published in 1975–78, a Soviet investigator, A. G. Maleeva, was able to compare the structure of the third upper molar of a water vole (*Arvicola terrestris*) that lived in the territory beyond the southern Urals, from the late Pleistocene to our times. This comparison showed a certain change in the characteristics of the phene complexes that were found: the gradual disappearance of some typical phenes and the appearance of new ones.

Comparisons of the phene pools of various generations on a scale of only several generations have been presented in the works of a number of investigators working with various insects, especially drosophila. Classical works of scientists of the American school of Dobzhansky (Anderson et al. 1975) showed that in thirty years (hundreds of generations of drosophila), the frequency of occurrence of the main phenotypes and genotypes did not change, but the concentration of rare phenes fluctuated sharply. This is supported by data on spittlebugs (*Philaemus spumarius,*) published in 1960–78 in a series of works by V. E. Beregovoi. The concentration of principal morphotypes, or morphs (phene complexes), in a single population on the northern outskirts of Sverdlovsk (Central Urals) fluctuated insignificantly over four years (95–84%), but the concentrations of rare phenes dropped to zero and changed by several times.

By studying large numbers of long-lived large animals in museum series, it is possible to compare the phene pool of various generations (the age of each individual can now be determined precisely within a half year by using bone microstructure). Such a com-

---

* This situation, clear in general terms, may become more complicated by one condition: in directionality it is possible to detect the effect not of contemporary selection, but the manifestation of the results of selection acting many, many generations before, which has formulated the specific structural traits or reaction norms of organisms. These traits unavoidably determine certain possible directions of the future process of selection. Intensive analysis of the problem of directionality in evolution shows that this "aftereffect" of selection, fixed in the structure of each living being, is one of the widespread evolutionary phenomena.

parison usually shows the difference between generations in the concentration of various rare phenes.

The fluctuation in concentration of rare phenes shows the existence of a stabilizing selection for principal traits and properties. Stability in the direction of selection in a number of the examples studied is enormous. Thus, polymorphism for shell coloring has been preserved for 20,000 years in certain populations of the snail, *Cepaea nemoralis*, in England. This was established by analyzing layered Quaternary deposits containing the shells.

Of course, the investigator does not find stability of phenotype over a span of several generations in all cases. The frequency of almost all phenes also fluctuates sharply (for example, in some populations of aphis lions, *Chrysopa aspersa*, the common vole, *Microtus arvalis*, the house mouse, and others). More careful study disclosed that in these cases the investigator was not dealing with random samples from a population, but samples from separate demes. In demes, stability of phene concentration is, as a rule, rarely disrupted from generation to generation since demes exist a comparatively short time—several generations in all. Having obtained material from one and the same place, an investigator is often actually dealing with different demes. From this fact (change in phene frequency in demes) we may conclude that if we find sharp differences in comparing population samples, then it is necessary to confirm that we are not dealing only with different demes; that is, we must determine whether the given sample is representative enough to characterize the whole population.

Another explanation of quite large fluctuations in phene frequency in proximate generations may be weak selection pressure on the traits studied. Which specific case confronts the investigator (weak selection or nonrepresentative sample) can be established only by further analysis.

These are some of the ways of studying the effect of natural selection on a population phene pool. Isolation is another elementary evolutionary factor whose effect can be elucidated by studying the phene pool.

Gaps in the concentration of separate phenes can be assumed to be the effect of one of the forms of isolation, of which there are quite a number in nature. Which of these will be found in specific studies that explain a difference in phene frequency cannot be determined without special analysis. (It may be territorial-physical, or ecological, or ethological, or phenological, or some other form of isolation.) We can identify isolating barriers in a barn inhabited by mice by using phenetic methods, but we can say what specifically determines this isolation only by using data from ethology, ecology, etc.

I have already often spoken of phenes-markers that characterize a group of individuals on various scales. The minimal marker is individual. The development of techniques of sound recording has made it possible to use characteristics of sounds made by animals to identify individual animals under natural conditions. There is nothing surprising in this: we easily distinguish the voices of people we know and rarely make mistakes; this means that the voices differ, or are marked, in some substantial way. Modern technology makes the same thing possible with respect to birds.

American ornithologists (De Wolfe, Kaska, and Peyton 1974) carried out interesting work some years ago. Obtaining individual characteristics of songs (selections of phenes and their combinations) of Gambel's white-crowned sparrows, they traced the migrations of these birds from breeding grounds in Alaska to winter habitats in California. Without banding, without tagging, but instead according to the phenes of song, it was possible to find a bird where it was wintering among hundreds of others!

Let us consider the possibilities of studying the phene pool of another elementary evolutionary factor, the waves of life (see chapter 1).

The evolutionary concept "waves of life" denotes a sharp change in the concentration of rare mutations, and the inclusion of rare mutations and their combinations in the action of selection directly, and not in the cycle of automatically selected genes. As a result of the thesis waves, the concentration of rare mutations may increase under the influence of new directions of selection or, on the other hand, may decrease sharply, almost to the point of disappearing from the population phene pool. No population exists without surges in numbers, and any sufficiently long-term population study will inescapably come upon this clear phenomenon, which may be expressed in very different ways.

Unfortunately, thus far very few precise observations have been made of the change in phene pool of a population at equal stages of the manifestation of such waves. Here are a few known examples. Studying the coloring of muskrats, *Ondatra zibethica*, in the Lower Tungusk region in the period from 1964 to 1972, A. V. Komarov noted a drop in numbers (1968–69) and a period of great numbers (1971–72). During the depressed years, among adult individuals, there were a great number of animals with red tones in their coloring, and during the period of increased numbers, animals with gray coloring.

At the beginning of the 1930s, the well-known English investigators K. D. and E. B. Ford first began a detailed study of fluctuations in hereditarily determined discrete traits in natural populations of some butterflies. They and their successors obtained very interesting results that have even now not been explained. In some cases, changes in the number of spots on the butterfly wings corresponded to the sharp change in numbers, and in other cases, no such correlation could be found. Only the phenetic technique can detect changes in the genotypic composition of natural populations at various stages of the waves of life.

Thus, study of the phene pool gives the investigator interesting and sometimes unique possibilities of analyzing three elementary evolutionary factors: natural selection, isolation, and the waves of life, or generally speaking, both directional and stochastic processes in natural populations.

Describing the Gene Pool

In order to study the gene pool of a population (or another group of individuals), we must take into account the presence of phenes of one kind or another and determine the frequency with which they occur in the given assembly of individuals.

In studying the phene pool, the investigator must examine many phenes simultaneously. The simplest technique is simple listing in a column, as in Vavilov's tables. But such a record is inefficient: the designation of each phene must be repeated every time it is mentioned. Each phene can be coded with letters, numbers, or other arbitrary signs. Usually, each investigator codes phenes in one way or another, often unconsciously.

Recording traits (coding) may be positional or nonpositional. Positional coding is a record in which each trait occupies a precisely determined place. A classic example of positional coding is the dental formula of man (*Homo sapiens*):

$$\frac{2—1—2—3}{2—1—2—3} \quad 2 = 32.$$

This formula is decoded thus: incisors, 2; canines, 1; premolars, 2; molars, 3. Teeth of the upper jaw are indicated above the line; those of the lower jaw, below the line. The dental formula of hedgehogs of the genus *Erinaceus* (including the common European hedgehog) is

$$\frac{3—1—2—1—3}{2—1—1—1—3} \quad 2 = 32.$$

In addition to a difference in the total number of teeth, the dental formula of the hedgehog differs from the dental formula of man by the presence of a second group of premolars.

The position of numbers in the recording of the dental formula must not be changed since it would not be clear which number refers to which type of tooth.

It frequently happens, however, that we must deal with individual teeth, and not with the whole row of teeth in the upper or lower jaw. Then we use a semipositional recording. Every tooth is designated by the first letter of its Latin name, and its location in the upper or lower jaw is coded by the position of the number (superscript or subscript) with the letter: M^3—molar (molares), upper third, PM_2—premolar (praemolares), lower second, etc.

An example of nonpositional coding is found in the series of works of Larina, who studied the phenetics of the skull of mouselike rodents. Separate traits were designated as shown below.

Variations of the predental opening located in the lower jaw in front of the molars: A—one large opening; B—one large and one small opening; C—several small openings located in a row; C_1—three or four openings of average size; C_2—three openings in a triangle; D—one large opening in the region of the eye socket; D_{1-6}—different variants of location of the frontal opening.

In this recording system, nothing will change if the arbitrary designations are placed in a different order. Instead of letters, numbers or arbitrary signs can be used with equal success. Nothing will change in principle if a separate letter is used for any small variation of the frontal opening, and not, as has been suggested, a number with a letter. But it was precisely this double code for traits (numbers and letters) that was chosen. This is quite proper since there are objective features in man's perception of various symbols that dictate the selection of this method.

Why do we distinguish seven colors in the multicolored rainbow? Why do we distinguish seven notes in the wide musical range? It seems that this is not at all accidental: the human psychology of perception is such that the limits of optimal rapid perception range from five to nine different symbols, events, or elements, and the average number of consistently perceived elements is exactly seven (Zhamoida 1972). For this reason, when we code phenes in describing the phene pool, it is desirable to group them in sets not exceeding nine elements (numbers, letters, etc.).

For a more detailed characterization of each individual, Larina proposed describing the phenotype according to combinations of phenes of the right and left side. Therefore the record of a separate phene for an individual will be double: AA, AB, D_1, D_3, etc. The record of the entire phene configuration of the individual in a symbol representation will be a line of paired symbols, for example, AAD_2D_3aa or $ABDD_1ac$, etc. (figure 5.3) The terminology here is not yet fully developed, but we will obviously be able to define the correspondence of phenes of the right and left side of the body as being homophenic, and noncorrespondence as being heterophenic (in analogy with homo- and heterogeneous).

The problem of asymmetry (antisymmetry, disymmetry, etc.) of phene expression deserves most serious attention. Of course, it would be naive to connect the expression of phenes of each side of the body with the action of one of the parallel DNA strands of the genetic code. There is no such direct connection. But the fact of different manifestation of phenes in various parts of the body may be useful for very important directions in population and ontogenetic analysis. Problems of biological symmetry from the genetic position have been recently taken up again in the works of B. L. Astaurov (1974), and from the position of phenetics, in the works of V. M. Zakharov (1976, 1980).

Thus, the first step in describing any phene pool is to give symbolic designations to every phene and to indicate in what concentration it occurs in a given assembly of individuals (sometimes it is possible to use only extreme variants of concentration, its presence or absence in the given group).

Besides the described symbolic record, there is a whole series of methods for graphic description of the phene pool. Symbols can basically characterize the qualitative selection of phenes in a group. A quantitative characterization of phenes, the concentration of each of them, must be given in tables, or, what is considerably more vivid, in various kinds of graphs.

The first and simplest graph representation of the phene pool is an ordinary histogram. It can show the rate of occurrence of each phene. Here again a positional and nonpositional approach is possible: each phene may be given a specific place in the histogram and in other histograms containing the same selection of phenes for subsequent comparisons; comparing changes in concentration of phenes in different groups of individuals taken from the population (samples) is quite simple.

Or we can go a different way. If each column on the histogram is shaded in a certain way, then the relative location of the columns will be immaterial, and the main symbol of the phene will be one kind of shading or another. Usually the two methods are combined. Some phenes to which a reader should pay particular attention are stressed by shading while the positions of all phenes are preserved.

The following convenient device can be used in comparing histograms. One of the samples is chosen as the original, and the concentrations of phenes in it are placed in order of decreasing value. Since the distribution of columns in a histogram of a phene pool is completely arbitrary, then any kind of transpositions can be made, and they can be combined in any way. The placement of phenes in other samples, however, must precisely correspond with their placement in the first sample. In comparing such assemblies of individuals, it is easy to note both marks of similarity and marks of dissimilarity (figure 5.6).

An interesting method of graphic representation of the phene pool is the various sector graphs. In a typical variant of such a graph, the area of a circle is divided into a number of sectors corresponding to the number of phenes. The size of a sector corresponds to the proportion of each phene. It is possible to use not just a circle, but practically any geometric figure, such as a square or parallelepiped within which it is convenient to depict graphically the required proportion of area. An important limitation in the use of a sector graph is the need to compare phenes of any one group of traits that will total a 100 percent concentration. Such a graph can present the incidence of 30 percent black, 30 percent red, and 40 percent colorless individuals, but it cannot present the incidence of 90 percent red and 40 percent with spots on the back.

For a graphic expression of traits related to most groups of phenes, a variant of the sector graph called a "wind rose" can be used. This method consists of the following: A circle is divided into as many equal sectors as there are phenes. The concentration of each phene in percents is plotted along the axis of the sector (the whole axis equals 100 percent). Then combining all the points plotted on the axes, we get a geometric figure, the total configuration of which is a graphic characterization of the phene pool. This method is interesting in that it gives a generalized characterization of the phene pool for rapid comparison while at the same time allowing subsequent analysis of the extent (concentration) of separate phenes. The limitation of the method lies only in the number of phenes compared; if their number is greater than eight or ten, the separate projections on the wind rose become somewhat small and not easily seen.

A third variant of the sector graph is a circle divided into exactly equal parts (along the circumference) according to the number of phenes. The concentration of a phene in this case is indicated not by the size of the sector itself, but by the size of its axis. Such a graph

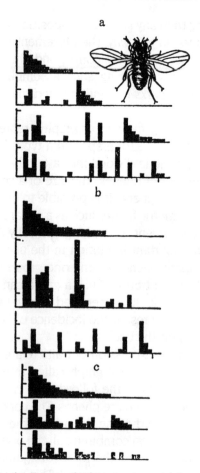

Figure 5.6. An example of using the histogram method to compare the phenetic composition of a population: a) percentage of various phenes ("mutations") in geographically separated (by hundreds of kilometers) populations of *Drosophila funebris*; b) the same in territorially adjacent populations; c) the same for one population over three consecutive years. (After N. V. Timofeeff-Ressovsky 1940)

represents something of an average between the usual sector graph of the first type and the wind rose.

We do not yet know all the methods of describing a phene pool—here we still have much work to do. For example, methods of machine analysis of the phene pool are already being tried. But all this is for the future. Even today, however, the investigator has at his disposal a whole complex of methods for describing the phene pool with which he can successfully resolve various phenetic problems.

Comparing Phene Pools

A precise comparison of phene pools is one of the most common problems in phenetics. In studying phene pools, it is necessary to compare them in some way in different populations, different groups of a population, different intrapopulation groups, and finally, a single population over time. In such a comparison it is of primary importance to determine whether the compared phene pools differ, and if they differ, then how and to what extent. Or conversely, in what ways are different phene pools similar? In general, we may say that qualitative comparison—presence or absence of separate phenes or their complexes, for example—tendencies toward change of the phene pools over large areas, can be done more simply with graphics using charts, conventional graphs, etc. (Many examples of such comparisons are presented in the following chapter). Sometimes a direct comparison on a graph, a chart, or a table is enough; the conclusion reached in this way may not need any special mathematical confirmation.

Until very recently no special statistical methods were developed for comparing populations and other intraspecific groups of individuals for frequency of discrete variations of coloring; such comparisons were made by using classical methods of comparing nonmetrical variations, such as Pearson's correspondence criterion (chi-square criterion), the Kolmogorov-Smirnov criterion (lambda criterion), or White's criterion. All these criteria make it possible to determine whether two compared groups belong to one general group or whether it is highly probable that they are from two independent, genetically isolated, samples (Penrose 1954).

To determine the degrees of similarity of several groups being compared, we can compute the coefficients of phenetic similarity, usually based on a computation of the so-called genetic distance (Mahalanobis 1936; Nei 1972). Recently another method of comparing

populations was developed that makes it possible to avoid complicated computations involved in determining the Mahalonobis distance and specially devised for comparing phenetic data (Zhivotovskii
1982).

I will designate the frequency of different variations in the
sample from one population as $p_1, p_2, \ldots p_m$, and the sample from
another population as $g_1, g_2, \ldots g_m$. The index of similarity of population
r will be computed according to the formula

$$r = \sqrt{p_1 g_1} + \sqrt{p_2 g_2} + \cdots + \sqrt{p_m g_m}.$$

The value of this index of similarity is equal to 1 when the
populations are identical with respect to the compared frequencies of
variation and to 0 when the compared samples do not have a single
common variation.

The statistical error, s_r, of the index of similarity for cases
where all the variations are represented in both samples, and the
samples differ only in frequencies, is determined according to the
formula

$$s_r \approx 1/2 \sqrt{\frac{N_1 + N_2}{N_1 N_2} (1 - r^2)},$$

where N_1 and N_2 are the size of the samples.

If several traits (variations) are being studied, then we may
determine the average similarity of the populations in all the traits as
an arithmetical average of the indices of similarity in separate traits:

$$r = 1/(r_1 + r_2 + \ldots r_n), \qquad s_r \approx 1/n \sqrt{s_1^2 + s_2^2 \ldots s_n^2},$$

where $r_1, r_2 \ldots r_n$ and $s_1, s_2 \ldots s_n$ are the values of indices of similarity
and their errors for corresponding traits, and n is the number of
compared traits. On the basis of the indices obtained in comparing a
large number of samples, we can construct a dendrogram of similarity.

It also is possible to compute a single index of similarity.
To do this, we must compute the frequency of morphs averaged for all
K populations. Then for each sample, similarity r to this "averaged"
sample must be computed. The index of similarity obtained in this
way can be used not only in comparing groups within a species, but

also in comparing any populations, including those of different species.

In analyzing phenes in populations and other groups of individuals, we can use the classical genetic method of investigation, comparison of distribution of the trait found under natural conditions with the Hardy-Weinberg equation, which has been described.

In 1908, the founders of mathematical genetics, the English mathematician G. Hardy and the German doctor W. Weinberg, simultaneously and independently of each other demonstrated that without pressure from any external factors, the frequency of genes in an infinitely large, completely panmictic population will already be stabilized after a single generation. It is true that such populations, infinitely large, panmictic, and without pressure from external factors, do not exist in nature. But just as a model of an ideal gas allows physicists to visualize what takes place in real gases, so the Hardy-Weinberg concept of an ideal population can find wide application in modern genetics for general evaluations of the concentration of individual genes and heterozygotes in natural populations. In phenetics, the Hardy-Weinberg formula can be used successfully to determine the incidence of two alternative phenes by comparing the empirical distribution with the theoretically expected distribution.

This can be done if extreme variants (preferably homozygous) and one intermediate (preferably heterozygous) class of individuals can be isolated from the material. According to the Hardy-Weinberg equilibrium, the frequency of alleles must be distributed according to the formula of Newton's binomial, and the frequencies of phenotypes and genotypes can be computed from this.

Let us examine a specific example from the study of a house mouse population caught at six farms not far from each other in Michigan (acccording to data from Petras 1967) (See table 5.3).

From a comparison of the data obtained with the theoretically expected, the predominance of the heterozygous phenotype II is obvious. Consideration of a number of possible explanations for the predominance of heterozygotes led the author to the conclusion that it was linked to a disturbance of panmixis. The groups of individuals examined belonged not to a single panmictic population, but to separate, much more inbred groups (breeding "within themselves").

Table 5.3 Comparison of Observed and Theoretically Expected Numbers of Esterase Phenotypes Observed in Mice

		Phene Frequency					
		Phenotype I		Phenotype II		Phenotype III	
Group	Number	Observed	Expected	Observed	Expected	Observed	Expected
I	76	38	36.96	30	32.08	8	6.96
II	109	62	57.98	35	43.03	12	7.98
III	40	18	17.56	17	17.89	5	4.56
VI	24	12	10.67	8	10.67	4	2.67
V	26	12	9.24	7	12.52	7	4.24
VI	21	14	12.96	5	7.07	2	0.69
Total	296	156	144.76	102	124.48	38	26.76

Further computations and ecological data confirmed the correctness of this hypothesis and even made it possible to determine the size of demes, panmictic units. In this study demes fluctuated from six to eighty individuals.

In using the device of comparing empirical values with data computed on the basis of the hypothesis on equilibrium of the Hardy-Weinberg type in the populations, we must not be carried away by the apparent precision of mathematical computations.

Sometimes it is said that in biology mathematics should be used as widely as possible. This position, in its general form, is improper. Sometimes we obtain figures that seem exact, but hidden behind them is the imprecision of biological samples; these figures essentially express nothing. No matter with what kind of astounding precision computations are made, the accuracy of investigations will not be increased if samples are imprecise. The results of such investigations usually place before the scientist an equation with many unknowns. It can be resolved correctly only by careful biological analysis of natural material. But the question of the prevalence of any of the observed forms can be raised on the basis of the Hardy-Weinberg formula.

Methods of generalized comparison of phene pools are used, one of which is comparison of intensity. This method was used especially in a whole series of interesting works on the phenetics of house and field sparrows, genus *Passez*, in the Mediterranean area by the German investigator W. Meise in 1936, and in simultaneously published works on the phenetics of *Tradescantia* by the American biologists E. Anderson and K. Sax (1936). The essence of the method is as follows: Variations of a separate trait (both discrete and nondiscrete variations can be used) are designated by a specific number of indices. The sum of the indices for all traits studied will give the generalized characterization of an individual, and the average number of indices for the group of individuals, the characterization of the whole group. The use of such generalized characterization is particularly successful in isolating hybrid forms in natural populations and describing graded variability (see below). Thus, the typical intermediate, hybrid form between the field and house sparrows was the Italian sparrow that has even been classified as a separate species (Johnson 1969).

The use of total characterization of any phenetic trait, or group of traits, is linked to the loss of some information. In my study of the structure of a population of bank voles, *Clethrionomys glareolus*, in the vicinity of Moscow (Krylov and Yablokov 1972), individuals were compared according to the average number of openings in the upper part of the eye socket, the alveoli of molars in the lower jaw, and the number of predental openings; that is, the comparison was actually done according to correlated indices. Studying the phenetics of the same species in the flood plain of the Ural River, V. N. Bol'shakov and A. G. Vasil'ev (1975) demonstrated that in the material they examined the groups of individuals that did not differ accroding to the average number of these openings differed distinctly in a comparison of frequency of independent variants. Obviously, such a comparison may yield a more detailed characterization than a comparison according to correlated summarized indices.

The selection of a method of investigation depends on the purpose: sometimes particular information is consciously given up in favor of the possibility of detecting general patterns.

In studying the phene pool, we must try to use an adeguately large number of phenes. An important problem is the selection of phenes optimal for the specific study (in scale, in resolving power of the experiment, and with respect to the inclinations of the investigator in choosing the method to study variability.) It often happens that intrapopulation units or the effect of evolutionary factors cannot

be detected in the available material according to preliminarily identified phenes. This does not mean that these phenes have been identified incorrectly; it may be that they appear on a different scale, larger
or smaller than the scale on which the study is being conducted.

At the present stage, actually the initial stage of the development of phenetics, the analysis of the phene pool is one of the
main directions in which the study of separate populations is going.
This kind of research may be especially productive in a sufficiently
long-term study of natural groups, over the course of a number of
generations. Here possibilities that have not yet been assessed are
opening up. The study of numerous remains of organisms makes it
possible—for many groups of small mammals, molluscs, and the most
various groups of plants and a number of other organisms—to include
in these investigations the Quaternary and Pleistocene periods, using
phenetic methods for analyzing paleontological material.

There is great interest in obtaining adequately precise
phenetic characterizations of populations, as if "in stock," for subsequent comparison over several years of transmission in the scientific
relay to the coming generation of investigators. Naturally, in these
cases, studies must be done on species that will not be extremely rare
or become extinct in the foreseeable future.

I would like to emphasize once more one of the dangers
in the organization of long-term studies of a phene pool—the danger
of developing a characterization on the basis of somewhat small and
inadequately representative material that will show not the frequency
of phenes in a population, but only the ephemeral variations of phenes
in demes. Isolation of phenes sharply fluctuating in frequency of
concentration in addition to consideration of an adequately large
number of individuals will help avoid this danger.

REFERENCES

Anderson, W., T. Dobzhansky, O. Pavlovsky, J. Powell, and D. Yardley. 1975. Genetics
    of natural populations XLII: Three decades of genetic change in Drosophila
    pseudoobscura. Evolution 29(1):24–36.

Anderson, E. and K. Sax. 1936. A cytological monograph of the American species of *Tradescantia*. *Bot. Gaz.* 97:433–476.

Astaurov, B. L. 1974. A study of hereditary disturbances in bilateral symmetry related to variability in identical structures within the organism. In *Heredity and Development*, pp 54–109. (In Russian.) Moscow: Nauka.

Bel'kovich, V. M. and A. V. Yablokov. 1965. The structure of a herd of toothed whales. In *Marine Mammals*, pp. 65–69. (In Russian.) Moscow: Nauka.

Beregovoi, V. E. 1966. Variability in natural populations of the common spittlebug (*Philaenus spumarius* L., Homoptera). (In Russian; English summary.) *Genetika* 11:134–144.

Bol'shakov, V. N. and A. G. Vasil'ev. 1975. Spatial structure and variability of populations of bank voles at the southern boundary of their habitat. In *Population Variability of Animals*, pp. 3–31. (In Russian.) Sverdlovsk.

DeWolfe, B., D. D. Kaska, and L. J. Peyton. 1974. Prominent variation in the songs of Gambel's white-crowned sparrows. *Bird Band.* 45(3):224–252.

Evans, M. 1977. Recognizing individual Bewick's swans by bill pattern. *Wildfowl* 28:153–351.

Evans, M., A. V. Yablokov, and A. E. Bowles. 1982. Geographic variation in color pattern of killer whales (*Orcinus orca*). *Rep. Int. Whal. Comm.* 32:687–694.

Ford, H. B. and E. B. Ford. 1930. Fluctuation in numbers and its influence on variation in *Melitaes aurinia*. *Trans. Roy. Entom. Soc.* (London), 78:345–351.

Ishchenko, V. G. 1971. Using morphological polymorphism in population ecology of mammals. (In Russian.) *Ekologiya* 5:64–70.

Ishchenko, V. G. 1978. *Dynamic Polymorphism of Brown Frogs within the Fauna of the* USSR. (In Russian.) Moscow: Nauka.

Johnson, R. F. 1969. Taxonomy of house sparrows and their allies in the Mediterranean basin. *Condor* 71(2):129–139.

Krylov, D. G. and A. V. Yablokov. 1972. Epigenetic polymorphism in the skull structure of the bank vole (*Clethrionomys glareolus*). (In Russian; English summary.) *Zool. Zh.* 51(4):576–584.

Larina, N. I., V. L. Golikova, and I. V. Eremina. 1976. The use of certain phenetic methods in the study of intrapopulation groups in mice and voles. In *Physiological and Populational Ecology of Animals* 4(6):69–79. (In Russian.) Saratov: Saratov University Press.

Mahalanobis, P. C. 1936. On the generalized distance in statistics. *Proc. Nat. Inst. Sci. India* 2(1):49–55.

Meise, W. 1936. Zur Systematik und Verbvreitungsgeschicthe der Haus- und Weidensperlinge, *Passer domesticus* und *hispaniolensis*. J. *Ornithol.* 84:631–672.

Nei, M. 1972. Genetic distance between populations. *Amer. Natur.* 106:283–291.

Penrose, L. S. 1954. Distance, size, and shape. *Ann. Eugen.* 18:337–343.

Petras, M. L. 1967. Studies of natural populations of *Mus*. II: Polymorphism at the T-locus. *Evolution* 21(3):466–478.

Smith, M. L. and J. Carmon. 1972. Pelage color polymorphism in *Peromyscus polionotus*. J. *Mamm.* 53(4):824–833.

Timofeeff-Ressovsky, N. V. 1940. Mutations and geographical variations. In J. Huxley, ed., *The New Systematics*, pp. 73–136. Oxford: Oxford University Press.

Turutina, L. V. 1982. A study of the spatial-genetic intrapopulation structure of two species of vertebrate animals (*Lacerta agilis* L., *Celthrionomys glareolus* Schreb.) using the phenetic method. In A. V. Yablokov, ed., *Phenetics of Populations*, pp. 174–187. (In Russian.) Moscow: Nauka.

Yablokov, A. V., A. S. Baranov, and A. S. Rozanov. 1980. Population structure, geographic variation, and microphylogenesis of the sand lizard (*Lacertra agilis*). In M. K. Hecht, W. C. Steere, B. Wallace, eds., *Evolutionary Biology*, 12:91–127.

Zhamoida, A. I., ed. 1972. *Cipher Coding of Systemic Traits of Ancient Oraganisms.* (In Russian.) Moscow: Nauka.

Zhivotovskii, L. A. 1982. Indices of population variability according to polymorphic traits. In A. V. Yablokov, ed., *Phenetics of Populations*, pp. 38–44. (In Russian.) Moscow: Nauka.

FOR ADDITIONAL READING

Brown, L. I. 1965. Selection in a population of house mice containing mutant individuals. *J. Mamm.* 46(3):461–465.

Jones, J. S. and A. J. Irving. 1975. Gene frequencies, genetic background, and environment in Pyrenean populations of *Cepaea nemoralis* (L.). *Biol. J. Linn. Soc.* 7(4):249–259.

Larina, N. I. and I. V. Eremina. 1982. Certain aspects of the study of phene and gene pools of a species and of intraspecific groups. In A. V. Yablokov, ed., *Phenetics of Populations*, pp. 56–68. (In Russian.) Moscow: Nauka.

Maleeva, A. G. 1975. A comparison of morphotypical variability in teeth of *Avricola terrestris* L. and *Lagurus lagurus* Pall. (Rodentia). In *Fauna of the Urals and Northern Europe*, 4:42–49. (In Russian.) Sverdlovsk.

Selander, R. K. 1970. Behavior and genetic variation in natural populations. *Amer. Zool.* 10(1):53–66.

Yudin, B. S. 1977. Individual and population variability in the dental system of the Siberian mole. In *Fauna and Systematics of Siberian Vertebrates*, pp. 178–199. (In Russian.) Moscow: Nauka.

Zakharov, V. M. 1982. The phenogenetic aspect of studying natural populations. In A. V. Yablokov, ed., *Phenetics of Populations*, pp. 45–55. Moscow: Nauka.

Zhivotovskii, L. A. 1980. Index of intrapopulation diversity. (In Russian; English summary.) *Zh. Obshch. Biol.* 41(6):828–836.

CHAPTER SIX

Phenogeography

Phenogeography is the analysis of the geographic distribution of individual traits—phenes and phene complexes, as a rule—within the limits of a species habitat, carried out while studying problems of microevolution, intraspecific systematics, and the development of biotechnical practices. Phenogeography is, in a certain sense, the quintessence of phenetics: a significant part of phenetic studies is directed toward obtaining phenogeographic data.

As early as the 1920s, in Moscow and Petrograd discussions took place on the need to study phene pools of those species that are of practical interest to man (such as domesticated animals, cultivated plants, hunting-trade animals, birds, fish, many forest species, and field plants of wild flora). At that time, the method for establishing species gene pools was to study the geographic distribution of the greatest number of different simple hereditary traits. Thus arose the study of geographic centers of morphopysiological diversity in a number of cultivated plants, later proposed by Vavilov as the basis for the theory of centers of diversity and origin of cultivated plants (see chapter 3). In addition, the formulation of a new direction, "genogeography," was proposed by Serebrovskii in 1928.

The purpose of genogeography was the geographically based desciption of the gene pool of hereditary variations within a species range, that is, producing geographic maps of the distribution of the frequencies of the greatest number of hereditary traits within habitat limits of a species. In the 1920s, among domesticated animals, the most studied genetically were domestic chickens, and to a lesser degree, cattle and horses, and among plants, certain cultivated grains and legumes. It was these groups especially that served as the first material of phenogeographic studies.

Vavilov and his coworkers began a study, impressive in scale, of the geographic distribution of various traits in grains and legumes, and Serebrovskii and his colleagues, a study of the gene geography of domesticated chickens. At the same time, under the direction of Vavilov, Filipchenko, Kol'tsov, and Serebrovskii, a collection of data on local species of cattle and horses was organized according to gene geography. The work was done partly in Kazakhstan with the participation of Dobzhansky.

Subsequently, interest in this research area dimmed (the main reason was, evidently, that unresolved problems in genetics distracted the investigators from the broad study of populations under natural conditions), and in the 1930s and 1940s only isolated works pertaining to the "distribution of mutations" in natural popualtions appeared. Some of these works still have significance, as, for example, the works of the German biologist K. Zimmerman (1935–39) on the distribution of the simplex mutation in the molars of the common vole in Europe, which became classics. Figure 6.1 shows the regular decrease in concentration of this trait in central Europe.

Interest in phenogeographic studies has surged again in the last fifteen to twenty years after the development of the theory of population genetics, as a result of biochemical methods of study and an avalanchelike increase in the number of studies of nonmetrical variations in natural populations.

Regardless of the now very significant number of works in this area being done abroad, the concept of phenogeography was developed most systematically in the works of V. E. Beregovoi (1965–1976), done mainly on a genus of wagtails (figure 6.2) and on some reptiles and insects.

Phenogeographic methods permit the resolution of the most complex problems in the study of intraspecific variability: identifying populations and groups of populations and determining population boundaries. Another important problem for phenogeography is the study of the effect of natural selection and other evolutionary factors—primarily, isolation. Finding the centers of species diversity within a species and reconstructing the historical development of

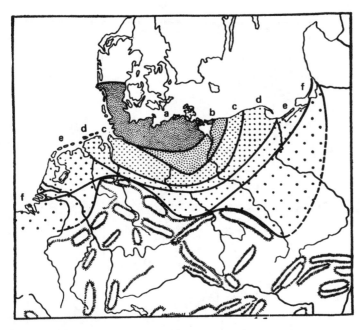

Figure 6.1. Distribution and concentration of a phene "simplex" of tooth structure in the common vole *Microtus arvolis* in central Europe: a) 90 percent; b) 70 percent; c) 50 percent; d) 30 percent; e) 15 percent; f) less than 15 percent. (Zimmerman 1935)

separate parts of the species population and the species as a whole (microphylogenesis) are also interesting problems of phenogeography. Besides resolving such evolutionary-theoretical problems, phenogeography serves as an important instrument in the study of intraspecific systematics and yields data for the proper organization of a number of managed fish and game practices. Let us examine all these directions of phenogeographic investigation.

Finding the Boundaries of a Population

Populations within any species differ from each other according to the frequency of occurrence of various alleles, which must be expressed externally in different concentrations of various phenes. For this rea-

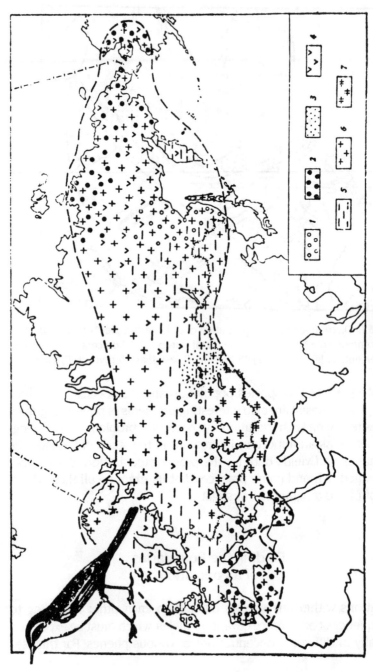

Figure 6.2. An example of phenogeographic investigation: phenogeography of the yellow wagtail *Motacilla flava*; 1) yellow head color; 2) olive-brown head color; 3) bluish shading of head feathers; 4) white brow; 5) blue-gray crown; 6) lead-gray crown; 7) black crown. (Drawn by V. E. Beregovci from Timofeeff-Ressovsky, Yablokov, and Glotov 1973)

son, if in studying a species population under natural conditions, we find a sharp break in the frequency of any phenes, we have reason to think that a population boundary exists at this point.

Let us consider a number of typical examples.

Figure 6.3 presents an example of the first goal-directed phenogeographic work: the concentration of phenes in a chicken population in Dagestan North Caucasus studied by Serebrovskii (1927). During the study, chickens lived near mountain villages in a semiwild state, nesting in bushes; for this reason, they could be studied as models of true natural populations. Since the genetics of basic traits in chickens was already known at that time, Serebrovskii used the results of such studies and transferred them to observations under natural conditions, isolating phenes.

The research was done in the vicinity of the river Avarskoe Koisu. Dagestan is a country of gorges, sometimes hundreds of meters deep, and Avarskoe Koisu flows in one such gorge. The width of the gorge is several hundred meters, but the chickens cannot fly over it; this provides a serious isolating barrier. Isolation among other groups of chickens studied was less significant. Groups of chickens living on one side of the river were similar in allele complement, but differed sharply in this respect from chickens living on the other side.

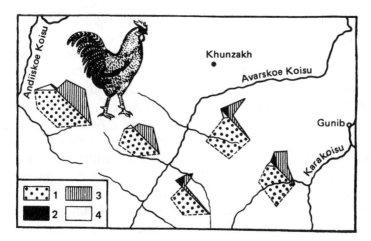

Figure 6.3. Phenogeography of chickens in Upper Dagestan: 1–4-phenes of color and shape of comb. (Serebrovskii 1927)

The width of population boundaries depends, although not directly, on the degree of mobility of the organisms. As a rule, in more mobile animals it is greater than in the less mobile—for example, in molluscs. In England, in a number of cases studied, the frequency of the phene for striping and shell color of the snail *Cepaea nemoralis* changes sharply at a distance of no more than 20–30 meters on a continuous area inhabited by this species (Jones 1973).

Figure 6.4 shows the concentration of the "red tail" phene in a population of common squirrels (*Sciurus vulgaris*) in the region of the upper Volga. As the studies of I. S. Tomashevskii have shown, over a distance of several dozen kilometers the frequency of this phene changes greatly, marking in this way the real natural boundary between populations. It is possible that this is a boundary not of one population, but of the whole group of populations since the boundary between subspecies of squirrels occurs precisely here.

Many examples of identifying population boundaries are provided at present by the study of biochemical phenotypes, the frequency of occurrence of various isozymes. A classical example is that of the well-identified natural boundaries between groups of populations of house mice in Jutland, Denmark. Data on the two different phenes, esterase 1 and 2, correspond, although the hybrid zone in each case (with respect to both esterases) has a specific configuration and latitude. Such a situation is very typical. Thus, for example, the hybrid zone between two populations of lizards (*Cnemidoforus tigris*) that belong to different subspecies in the southwestern part of New Mexico, determined according to the morphological traits of scaliness and coloring, is approximately 2 km, and analyzed according to biochemical phenes, exceeds 40 km (Zweifel 1962; Dessauer, Fox and Pough 1962).

The common two-spotted ladybug, *Adalia bipunctata*, provides an interesting example of determining population boundaries. It is found in two basic forms, black and red. Different populations and groups of populations within the species are distinguished according to the frequency of these forms. In 1976–77, I. A. Zakharov and S. O. Sergievskii (1978) observed that one of the boundaries between populations, determined according to the change in distribution of these forms, passes along the central regions of Leningrad for a distance of several kilometers.

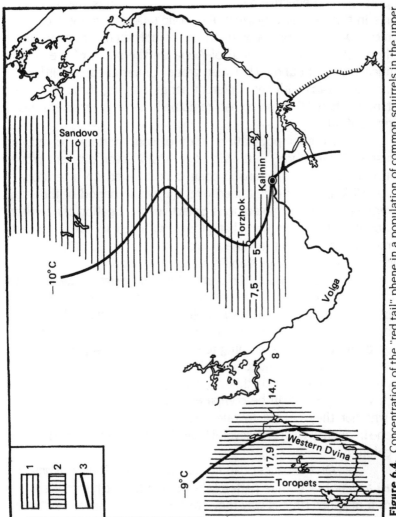

Figure 6.4. Concentration of the "red tail" phene in a population of common squirrels in the upper reaches of the Volga River: 1) northern population; 2) western population; 3) January isotherm. (Tomashevskii 1969)

As a final example, I will cite two cases of identifying populations boundaries in nature according to phenetic methods; these involve the character of bird song. Now, because of the wide use of portable and effective methods of recording songs, the number of works in this area is very great. Parabolic reflectors make it possible to record with great precision the voices of individual birds at distances of hundreds of meters. South of Accra on the shore of the Gulf of Guinea, a sharp boundary between groups of nectar-eating birds (*Nectarina coccinigaster*) with very different songs, passing through an area not more than 50 m wide, has remained stable over several years (Grimes 1974).

Another population study of song pertains to redwing thrushes (*Turdus iliacus*) in the vicinity of Oslo. Seven different groups of thrushes with characteristic song dialects lived on a territory of about 85 km, stably maintained over the whole period of the study covering two or three generations. Selective investigation of a territory of about 250 km^2 showed that groups with ten to twelve dialects formed a "superdialect," a large group of related dialects (Bjorke 1974), and constituted, in all probability, a true population.

Two basically different situations exist in determining population boundaries under natural conditions. Boundaries between populations may be sharp, easily determined, or they may be unclear, diffuse, in which case the populations may be united by a whole range of gradual transitions according to concentration of separate phenes. In the latter case, phenogeographic methods help identify only the population centers. With this approach, randomly chosen samples from various areas can be compared, and a determination made as to whether they belong to one common population, according to phene frequency. I will give several typical examples involving the study of coloring.

On the backs of adult male harp seals (*Pagophilus groenlandicus*) a clear black-white pattern always occurs that resembles wings from a distance (this is why they are called the "winged ones" along the shore). The pattern invariably differs in details. Our studies show that according to the frequency of basic types of patterns (figure 6.5), the group of harp seals living in the waters off Newfoundland differs

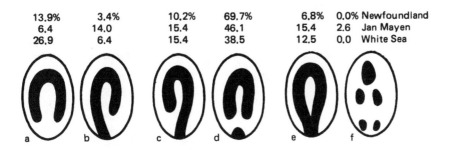

13.9%	3.4%	10.2%	69.7%	6.8%	0.0% Newfoundland
6.4	14.0	15.4	46.1	15.4	2.6 Jan Mayen
26.9	6.4	15.4	38.5	12.5	0.0 White Sea

Figure 6.5. Different types of coloring (a–f) of male harp seals, the concentrations of which indicate statistically reliable differences between the populations of Newfoundland, Jan Mayen Island, and the White Sea. (Yablokov and Etin' 1965)

from those reproducing in the waters of the Greenland Sea (in the region of Jan Mayen Island) and in the White Sea. Such statistically reliable differences in frequency of various types of coloring served as one of the arguments in favor of genetic independence of the groups studied, of their belonging to different populations (which was subsequently definitely confirmed by skull differences and tagging). Another example: According to the frequency of a white spot on the breast among the 1183 cats examined in the streets of Paris (Dreux 1978), it was found that at least three populations of cats live there in a wild state (the genetic dependence of this trait of coloring has been known for a long time). Many examples of this kind could be cited.

There is no other being whose body form has been studied as much as that of man. In ethnic anthropology, there are literally hundreds, if not thousands, of examples of differences between separate populations of people according to various genetically simple traits (phenes).

Table 6.1 presents examples pertaining to phenes, the study of which is practically impossible in the world of animals: the ability to taste phenylthiocarbamide (some people do not taste it at all; others find it bitter, like quinine); color blindness in which people

Table 6.1 Phenetic Characteristics (frequency of the trait in percent) of Certain Human Populations

Population	Ability to Taste Phenylthio-carbamide	Population	Color Blindness	Population	Number of Dizygotic Twins per 100 Births
Hindus	33.7	Belgians	8.6	Negroes (Nigeria)	39.9
Danes	32.7	Hindus	8.1	Negroes (Zaire)	18.7
English	31.5	Norwegians	8.0	Greeks	10.9
Spaniards	25.6	Scots	7.8	English	8.9
Malaysians	16.0	Germans	7.5	Swedes	8.6
Japanese	7.1	Chinese	6.2	Italians	8.6
Laplanders	6.4	Mexicans	2.3	French	7.1
Chinese	2.0	Indians (USA)	2.0	Spaniards	5.9
Indians (Brazil)	1.2	Negroes (Zaire)	1.7	Japanese	2.7

SOURCE: Jorgensen et al., 1964.

Table 6.2 Phenetic Characteristics (frequency of traits in percent) of Several Tribes of the Indigenous Population of Northeastern Afghanistan and Eastern Punjab

Trait	Kafir (Afghanistan)				Punjab	
	Paruni	Bashgali	Kalakh	Chitrali	Korzok	Yangpu
Eye opening:						
Round	0	4.0	79.5	76.5	0.0	0.0
Oval	91	85.2	19.8	23.5	4.6	82.5
Semioval	9	9.5	0.6	0.8	36.0	17.5
Slitlike	0	1.3	0.0	0.0	59.4	0.0
Nostrils:						
Concave	6	5.3	4.8	2.4	13.8	0.0
Straight	64	40.0	13.8	16.0	64.0	49.6
Curving	6	14.7	17.4	7.2	9.2	4.2
Convex	24	40.0	64.0	74.4	13.8	46.2

SOURCE: Jorgensen et al. 1965.

afflicted by daltonism do not distinguish red and green; and the frequencies of birth of dizygotic twins. Table 6.2 presents examples of differences according to fine features of facial structure (from Jorgensen et al. 1965).

The next two examples in this series pertain to the study of behavior. Individual variability and geographical differences in bird song, which we have already mentioned above, were known to naturalists, both professional and amateur, for a very long time. Differences can be distinguished between Kursk and Kiev nightingales, and Thuringian and Frankfurt finches; many other examples were widely recognized as early as the eighteenth and nineteenth centuries.

In the central region of the European part of the USSR, the chaffinch, *Fringilla coelelis*, with its relatively simple song, is suitable for such studies. In one of his works, A. N. Promptov (1930) compared the songs of chaffinches from two areas, the vicinity of Moscow and the western Urals. On the basis of the analysis of several thousand songs, he established that in the vicinity of Zvenigorod (70 km from Moscow) there were more individuals with two- and four-figure songs, while singers with three-figure songs were more likely to be found near Moscow. No four-figure singers lived in the western Urals. At the same time, in the Urals much more often than in the Moscow area, whistling variants occurred in the second figure of the song; near Moscow, crackling sounds predominated in this figure. Finally, in the Urals, single-figure song variants were found that never appeared in the vicinity of Moscow. The exceptional simplicity of such investigation is attractive: there is no need for complicated recording and decoding apparatus; a notebook and pencil suffice. Data on the geographical variability of song elements are now available for almost all the birds of Europe and North America. A real advantage of the cases described is that there is no need to kill the animals to obtain data on the phene pool and phenogeography.

All these examples demonstrate two main methods of identifying populations under natural conditions: finding sharp boundaries marked by notable breaks in the distribution (concentrations) of phenes or comparing phenetic characteristics of separate groups of individuals in order to confirm whether they belong to different genetic groups. It is natural that for detailed investigation these two approaches can be combined—first in a general way, in order to identify populations in comparison with groups of individuals, and then in order to find a more clear natural boundary between them.

Is the Species More Variable
at the Center or the Periphery
of a Habitat?

For certain species that have been adequately studied phenetically, centers of species diversity can be identified in those parts of the habitat where the majority of the phenes examined are represented. Centers of intraspecific diversity were identified in Vavilov's works on the centers of the origin of cultivated plants, which have been mentioned several times. Such centers identified the most promising area for finding new forms for selection. On the basis of general evolutionary-genetic concepts, we can construct a hypothetical diagram of diversity of a species within the limits of the habitat.

As a rule, the center of phene diversity corresponds to the optimal zone of the species. This zone is characterized by a population comparatively stable in numbers and substantial in size. Here phene diversity, typical for the whole species, is maintained and is apparent. As distance from the center increases, the complement of optimal living conditions decreases. The populations on the periphery of the species habitat are usually small, subject to greater fluctuations in numbers and at a greater distance from each other; here selection vectors oscillate greatly. As a result, in such peripheral populations the phene pool will be impoverished (homozygotization of the allele pool). In every peripheral population, the phene pool will be poorer than in any of the central populations. In such extreme populations, rare phenes will be picked up more often, whereas they remain hidden in the heterozygotic state in the optimal zone. For this reason, the sum of the phene pools of peripheral populations may be more variable than that in the central populations. However, each of the extreme populations itself must be less variable phenetically than the central populations.

If the situation described is true, it leads to this conclusion: when, under natural conditions, an investigator finds a comparatively small population with sharp fluctuations in numbers and comparatively small phenetic variability in comparison with other populations, it may be assumed that he has found a periperal population. Isolated colonies of edible snails at the northermost boundary of the species habitat in Scotland were such peripheral populations (with decreased heterozygocity according to well-defined shell coloring phenes). Some of these populations were altogether monomorphic for separate phenes (Jones, Leithe, and Rawlings 1977).

This is essentially borne out by data in the detailed work of the Sibirsk zoologists B. S. Yudin (1972) and N. G. Shubin and M. L. Sedakova (1982) on tooth structure of the Siberian mole (*Ascioscalops* [*Talpa*] *altaica*). In the mountainous regions of southern Siberia, the oldest sections of the species habitat, deviations from the typical number of teeth were found in only 15.1–25.0 percent of the individuals, while at the periphery of the habitat, variability increased sharply (in the Tomsk region, up to 61.4 percent, and in the trans-Baikal area, up to 100 percent). It must be emphasized that, in complete agreement with the explanation of species variability presented above, the more rare deviations in the structure of the dental system were found not at the center, but at the periphery of the habitat.

Of course, the situation described above does not cover all the natural variability observed. The zone of the ecological optimum of the species may not agree with the geographical center of the habitat. An especially complex picture develops in mountain regions where the "ecological periphery" may deviate from the optimal zone by no more than several hundred meters. Under such conditions, the clearly expressed variability of physico-geographic conditions and the isolation of small populations related to this leads to an impoverishment of the phene pool (to strong homozygotization) even under relatively favorable living conditions.

Theoretically, it is clear that the phenetic variability of a population is linked to the variability of the natural conditions in which the population exists. But in some cases, it is not possible to identify any one center of variability on the basis of available phene samples. This is exactly what we found on the basis of an extensive examination of the sand lizard. In this species, within the limits of its habitat, there were, evidently, several centers of phenetic variability, and in this case we must speak of the *polycentrism* of the species.

For fish and game species, elucidation of the problem of

centers of variability has a practical significance; theoretically, this problem is interesting because it illuminates the basic ways species begin and develop—the paths of microphylogenesis.

Reconstruction of Microphylogenesis

Phenogeography makes it possible not only to determine population boundaries and identify population centers, but also to reconstruct the phylogenesis of a species (relative to the species scale of phylogenesis, more precisely called "microphylogenesis"). Keys for reconstructing microphylogenesis are the decoding of "stratified" individual phenes and their groups and identifying paths of possible migration marked again either by individual phenes or more often by their complexes.

I will illustrate the phenetic approach to reconstructing microphylogenesis with anthropological examples in which phenetic data have definite historical confirmation.

In Eurasia a definite boundary exists between populations with different (large and not very large) numbers of spatulate incisors. This trait constitutes a boundary between the Mongoloids, who have a large percentage of spatulate incisors, and the Negroid-Europeoid branches of development of the species *Homo sapiens*. It marks the historical processes as being no less than several tens of thousands of years old.

Interesting also are the exceptions to the general rule: occurrences of the Europeoid type of incisor correlation deep in the Mongoloid race habitat. In eastern Siberia, as the extensive studies by A. A. Zubov (1973) demonstrated, such exceptions involve the indigenous population of Olekminsk, Kirensk, and Vitim, the first Russian settlements made during the colonization of this region in the sixteenth century. But if we add certain other phenes of tooth structure to the frequency of spatulate incisors (such as, for example, Corabelli's cusp), we can even establish from which province of central Russia the first Russian settlers founding these towns in Siberia came!

In an analogous way, according to phenes marking a flow

of migrants, it is possible to reconstruct where a colony of Jews in India or a colony of Japanese living on the Amazon for more than century came from.

Data on concentration of spatulate incisors indicate that phene concentration cannot always and everywhere be decoded with adequate reliability. Diversity of phenetic composition with respect to spatulate incisors in Oceania requires special study. Did a destabilization occur here for unknown genetic reasons, and did this trait lose its importance as a racial trait, or was there a strong mixing in antiquity of various racial migration waves?

Ethnographers and anthropologists offer no single opinion as to how Oceania was settled. Some believe that settlers came from the American continent, while others maintain that they were of Asiatic origin. The frequency of spatulate incisors shows that there is reason for argument and that there will probably not be a single, unequivocal resolution. In this case the phenetic approach is good because it gives objective evidence of evolutionary-genetic processes which the investigator must decipher.

The following example will be of interest to those who like cats. The genetics of cats has been studied comparatively well, and at least nineteen different alleles are known that are marked by distinct phenes which can be recognized from a distance, including fifteen alleles of coloring and spotting, two alleles of fur quality (long-haired and short-haired), and two alleles of tail length. The concentration of these phenes (alleles) has been established in dozens of places. The spread of domestic cats over the earth during the period of major geographic discoveries originated basically in Europe. Being very well adapted to a semiwild existence with man (in the "shadow of man"), cats were subjected relatively little to selection and, as observations show, the frequency of phenes in their population is maintained quite stably and can serve as a good marker of the flow of genes. Thus, for example, a link has been established between the similarity in frequency of recorded phenotypes among populations of cats in Chicago, Saint Louis, and Lawrence and the direction of the main road used by the first English settlers who were going to the 'Wild West' of North America along the famous Santa Fe Trail at the end of the seventeenth and the beginning of the eighteenth centuries" (Todd, Glass, and Creel

1976). The frequency of certain traits in a population of cats in the cities of Texas demonstrates the merging of two phene pools of different origin—from the north, those coming with the English colonists, and from the south, those coming from Central America with the Spanish conquistadors.

The phenetics and phenogeography of cats also help to reconstruct the expansion routes of the Vikings. Not long ago, an enthusiastic investigator of the phenetics and genetics of cats, the American biologist N. B. Todd, demonstrated, on the basis of studying the phene frequency of aboriginal populations of cats, that the Vikings, in moving westward, colonized islands in this sequence: Shetland Islands, Orkney Islands, Outer Hebrides, the Isle of Man; then, somewhat later, the Faroe Islands and Iceland. As the cat phenes indicate, the Vikings came to these places directly from Scandinavia (Todd 1975).

Research on the structure of the skull of the Orkney vole (*Microtus arvalis orcadensis*) by the famous English geneticist R. J. Berry allows a look into still greater antiquity. Finding this vole on the Orkney Islands is in itself surprising: all of Great Britain is inhabited by another, closely related species, a British vole (*Microtus agrestis*). At the same time, traces of Orkney voles not less than 6000–7000 years old were found in the earliest layers in excavations of the last Stone Age. From ordinary zoogeographic reasoning, it seemed that the vole must have come to the Orkneys from the Shetland Islands, to which they had come with the Vikings from Scandinavia. But comparisons of the skull phenes of the Orkney voles and the Scandinavian voles showed significant differences. Unexpectedly, the greatest similarity was found with voles from the shores of the Adriatic Sea. This discovery pleased archeologists since they had long suspected the existence of direct connections between cultures of the Stone Age of northern Great Britain and the builders of the megalithic structures of the eastern Mediterranean. Thus even in that remote time, man together with a complement of fellow travelers in the form of voles could travel over the entire marine space of Oecumenia.

The next two examples also relate to man's mobility.

The first pertains to the geographical distribution in North America of certain traits of the snail *Cepaea nemoralis*. This species came into the New World from Europe only in the nineteeth century, and now covers vast areas right up to the Pacific coast. Study of the phenogeography (banding, shell coloring, and three biochemical phenes) showed that all North American populations can be divided into two unequal groups: a smaller, which includes the population of this species in Virginia, and a larger, which includes all other populations. Combining the phenes studied made it possible to formulate the hypothesis that the Virginia populations of the snail came from Italy, while all the other populations of this species in North America originated in northern European populations (Brussard 1975).

The second example also involves one of the settlers of North America, the corn borer (*Ostrinia nubilalis*), which is very damaging to agricultural plants in some places. One of the methods of combatting this pest is to attract the butterfly into a trap with pheromones. For two isomers of the strongest pheromone, ll-tetradecinylacetate, two "pheromone phenotypes" were found: some insects in the population were susceptible to the action of one isomer, others did not react to this isomer at all, but reacted to a different one. A study was made of twenty-eight European and fourteen American populations according to this trait (Klun et al. 1975). Exactly the same pheromone characteristics—in essence, physiological phenes—were found in populations from Pennsylvania and New York as in the populations near Bologna (Italy) and Wageningen (Netherlands). It seems that in 1909–14 large consignments of grain were shipped to North America from just these cities; the corn borer could have come in these and preserved its phene pool so stably for the last seventy years.

Such examples of decoding historic events by means of phenogeography point to the enormous possibilities of phenogeographic analysis in general.

Other examples involve no human activity—examples of natural evolutionary processes and their phenogeographic decoding.

Finnish investigators studied the distribution of four phenes of junctures of the prefrontal scales in a viviparous lizard (figure 6.6). In frequency of incidence of various phenes if turned out that Finnish populations are similar to the Karelian and east European, and the Swedish, to the central and west European. Such a

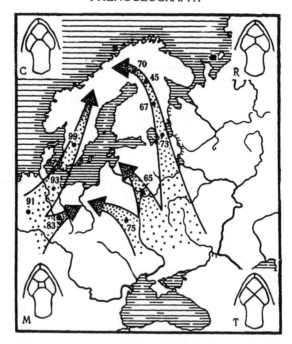

Figure 6.6. Differences in junctures (C, R, M, T) of prefrontal scales in viviparous lizards, *Lacerta vivipara*, and concentration of the "M" form in the European population of this species. Arrows indicate the assumed paths of migration; numbers indicte the concentration of the "M" form in percents. (From Voipio 1969, with new data from V. N. Orlovoy)

phenogeographic picture can be easily decoded in light of comparatively recent historical events. About 11,000 years ago, all of Finnoscandia was covered by a thick glacier; as the glacier melted and it became warmer, lizards gradually spread into this territory. Phenes indicate that the populations originated from two different sources—the first from west and central Europe through the then existing dryland neck at the Baltic Straits, and the second, from some centers in the southeastern part of Europe where this species had survived.

The next example is on a significantly smaller scale and pertains to a comparison of four bichemical phenes of several populations of old-field deermice (*Peromyscus polionotus*). The populations of this species that were studied live on islands in the Gulf of Mexico and on nearby parts of the mainland. On the islands, the phene pool was very homogeneous: only one phene appeared out of the four characteristic for this part of the habitat (Selander 1970). On the basis of these data, two hypotheses can be formulated. The first is that all four populations are genetically close to each other and all come from the same "root," a small group that came to the islands at some time.

Where did these original forms that arrived on the islands come from? Because of the similarity of the phene pool of the deer-mice, we may conclude that the population on the islands came not from the nearby shore populations, but from a population that is now quite distant from the shore, for which the predominance of this phene is characteristic. This is one possible answer, but there is another. The phene complement, or phene appearance, of the island populations and the nearby populations from which they were derived may differ because among the small number of individuals (founders of the island population) there might not have been representatives carrying the other three traits (the "founder principle" or "bottle-neck effect" are well known; these are only special cases of the effect of the waves of life). The second hypothesis is based on the fact that the unique phene appearance of the island populations may be favored by natural selection directed against the carriers of the other biochemical alleles, except for the single one that favors an increased viability under the given conditions.

Both these hypotheses are subject to further confirmation. The populations must be compared with respect to other phenes. Correspondence of one or two phenes may be accidental, but the probability of correspondence for several phenes is negligibly small and can be disregarded. Another method of confirmation is a comparison of the conditions of life of the island populations with the population distant from the shore, similar in biochemical characteristics, and any other populations on these and other islands where a similar direction to natural selection occurs. Thus from a equation that has several unknowns, we can gradually arrive at the desired equation with one or two unknowns.

An interesting example is research done on the Hawaiian-drosophila. (The American geneticist H. Carson [1970] describes it in "Chromosome Traces of the Origin of Species.") The frequency and location of bands (dark disks) in five chromosomes of sixty-nine species of Hawaiian drosophila were studied. These typical, discrete traits pertaining to fine chromosome morphology can be considered morphological phenes. They are good candidates for study because on the path from the gene to the trait, they lie "close" to the genes. All species examined were divided according to striation of chromosomes into three groups, each of which included phylogenetically close forms. These groups, judging from phenetics, come from an ancestral form living on Maui Island. Since the geological history of the separate islands has been quite well studied, it is possible in this case to reconstruct, in a way, the course of the evolutionary process over several hundreds of thousands of years. The method of comparison that the author used led to the hypothesis that several species of drosophila now living in the Hawaiian Islands come from a single female: this is borne out by the fact that all have a unique combination of discs that could not have developed independently in different individuals.

Undoubtedly the studies of the Hawaiian drosophila were influenced by the classical works on North American drosophila carried out by A. H. Sturtevant and Th. Dobzhansky in 1936. These works demonstrated that, knowing the order of genes in one of the chromosomes, one can reconstruct the order and sequence of origin of a series of sequential inversions (180 degree turns of segments within the chromosome), each of which is characteristic for discrete populations and groups of populations. This work remains the best in world literature as far as precision and unambiguity of results are concerned. It was done not on the phenetic, but on the genetic level: all stages were confirmed by numerous experiments interbreeding drosophila of various populations.

Although phenetic data taken separately are not as certain as genetic, in the aggregate they lead to well-founded microevolutionary conclusions.

The last of the series of examples illustrating the phenogeographic approach to decoding microphylogenesis concerns the study of a population of field mice, *Arodermus sylvaticus* in England. This, one of the most thorough phenogeographic investigations ever, was done by R. J. Berry, who studied twenty skull phenes of field mice living in Scotland and on nearby islands. The history of the settlement of these islands, which is divided into two periods, was established according to the summarized coefficient of coincidence. The first began at the time of maximum glaciation of Europe when only the southernmost part of what is now England remained free of ice (figure 6.7). It was precisely here that the then small population of mice survived

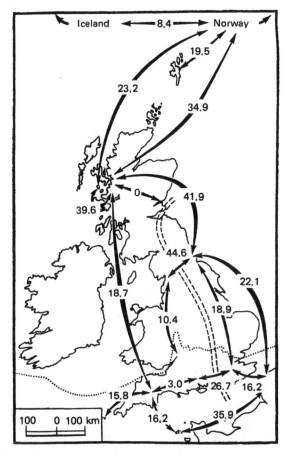

Figure 6.7. Phenetic distances between populations of field mice, *Apodemus sylvaticus*, in various areas of great Britain and neighboring countries, in arbitrary units. The dotted line indicates the boundary of maximum glaciation; the double dotted line, the modern habitat of preglacial populations. (Berry 1977)

the glacial period. As the glacier retreated to the north, the mice began to spread in that direction, but they occupied only the southwestern part of England.

Several thousand years after these events, the second historical period began. The Vikings arrived in the north of Scotland from Norway (their path can be traced through phene frequency in the cat population), and with them also came Norwegian field mice. As indicated by both archeological studies and phenetic comparisons, Little Eigg Island in the Inner Hebrides became a permanent center of Viking activity. From here, subsequently, the field mice spread over the whole northern part of England. Now field mice that originated in Norway live in most of the country; the descendants of the indigenous, preglacial population live only in the southeast.

The reconstruction of microphylogenesis of a large number of species makes it possible to obtain much interesting data on how the process of microevolution took place, and it will open new possibilities for a deeper understanding of the features of one set of species or another. Of course, such reconstruction will be especially convincing if an adequately large aggregate of phenes is considered.

Phenogeography,
Structure of Species,
and Systematics

Undoubtedly, phenogeography will also help resolve certain problems of interspecific taxonomy.

Formally, a subspecies is identified if no fewer than 75 percent of the individuals of a given group differ with regard to some traits from the rest of the species. Most often, subspecies are identified according to quantitative traits—intensity of coloring, size and proportions, differences in some ecologhical-physiological processes (reproduction times, for example)—connected with geographic location. The advantage remains, of course, with morphological traits since the systematicist wants to know to which subspecies this or that museum specimen belongs, when frequently nothing is known about it except its morphology and where it was collected.

The phenetic approach at a certain stage of argument about the reality and principles of identifying subspecies would "pour oil on the fire." What good, for instance, is the example of the salamander, *Plethodon jordanii*, in the Appalachians? The habitat of individual phenes within the species habitat has been identified (figure 6.8). At

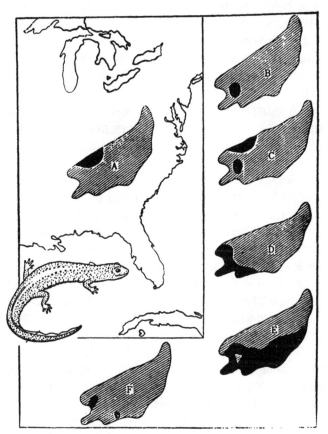

Figure 6.8. The phenetics of coloring of the salamander, *Plethodon jordani*, in the Appalachians of the United States. Black indicates the areas where 95 percent of the individuals have a) red cheeks; b) red legs; c) red spots on the backs of newly hatched individuals; d) white spots on the sides; e) dark abdomens; f) small, copper-colored spots on the back. (Highton 1962, from E. Mayr 1970.)

the same time, within the separate habitats, no fewer than 95 of 100 individuals have one of the indicated traits. Areas of discrete traits of coloring (phenes or their complexes) are distributed within the habitat of the species independently of each other and are combined in different variants. Formally, a separate subspecies of this species of salamander should be identified on the basis of each of these phenes. But in such a case, one and the same population will have to be classified as belonging to different subspecies (although in the area of every phene there must, of course, be many populations). Moreover, since the specified combination of phenes may also be considered as a taxonomic trait, should not a third subspecies be identified at points where two phenes overlap, and a fourth, where three phenes overlap, and so forth?

A subspecies must express real degrees of evolutionary-genetic divergence through which a species may pass in the process of evolution. Phenetics makes it possible to identify such steps of evolution objectively and not formally.

Subspecies must be identified as those historically com-pounded complexes of a population united by a common origin, a common adaptation to comparatively similar conditions (although, of course, microconditions of the life of each population will always be unique), and a common phene pool. Subspecies must reflect one of the higher levels of the intraspecific hierarchy. The duration of their existence on the average must be greater than the duration of the existence of separate populations, and the stability of their character-istics must also be greater than the stability of the characteristics of separate populations.

At this time there are more than a few examples demon-strating a surprising stability of population complexes. Even at the beginning of the 1950s, the French investigator M. Lamotte (1951), in a basic work on the study of phenes of shell striping in natural populations of the snail (*Cepaea nemoralis*), showed that variability is greatest in the smallest populations; in population complexes varia-bility is sharply curtailed. A number of such examples is collected in the works of Y. G. Rychkov (1969–1977). One such example—population structure and phene appearance of the indigenous Mongol population of northern Asia—illustrates the principal purpose of the investigations of this scientist. Over a period of years, Rychkov and his colleagues studied the distribution of biochemical phenes (blood groups, different electrophoretic variants of blood serum proteins, haptoglobins, transferrins), one physiological phene (sensitivity to phenylthiocarbamide: some people taste it as very bitter; for others it is tasteless), and on a somewhat smaller sample, morphological phenes of the structure of the skull.

On the whole, data were obtained on fifteen biochemical, twelve morphological, and one physiological phene for 212 populations. The investigations covered up to 2 percent of the indigenous population of nothern Asia and up to 88 percent of the existing ethnic groups. No other known study equaled this in degree of inclusion of a species population over such a wide part of its habitat. The results of such research are very significant for understanding geographic intraspecific variability. It was possible to compare all the main contemporary and ancestral populations of aborigines of northern Asia (neolithic, living in the same area 30,000 to 35,000 years ago). These groups are separated from each other by 200–280 generations. Regardless of the uniqueness of each separate population, the similarity of the general phene pool of the neolithic and contemporary population complexes is striking (figure 6.9). Processign of data for a number of other species (molluscs, fish, plants) made it possible to demonstrate that *populations within a species form a complex system—the more stable, the higher the hierarchic level of such a system.*

Another aspect of phenogeographical analysis concerns species structures, i.e. the greater similarity of populations that have a common origin and are, as a rule, spatially close to each other. Close populations are similar in a great number of traits, and different in comparatively rare and few traits.

This position on the closer phylogenetic relationship of spatially close populations must not be taken as an absolute. The process of evolution is quite multifaceted, so that sometimes territorially close groups of individuals may be branches of different microevolutionary processes within a species (for example, the case described above of the Orkney vole that reached the northern shore of Great Britain with the first sea voyagers of the Stone Age from the eastern Mediterranean, and not from areas nearby).

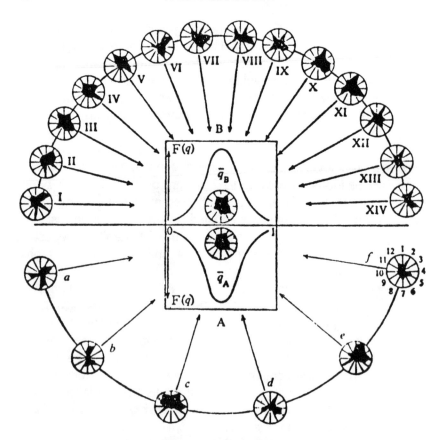

Figure 6.9. Comparison of incidence of frequency of twelve phenes of the structure of the human skull (1–12) in neolithic (a–f) and contemporary populations of northern Asia, separated in time by approximately 200 generations. (Rychkov and Sheremetyeva 1977)

In population and evolutionary biology, we frequently turn to results obtained with drosophila, genetically the best studied group of all the organisms that inhabit the earth. Especially pertinent are the results of one of the most exhaustive studies of comparative similarity of a population and population complexes of various hierarchic ranks. I am speaking of the investigation done under the direction of one of the students of Dobzhansky, the American geneticist F. J. Ayala.

There are fifteen closely related species in the group *Drosophila willistonii*, endemic to South and Central America. Biochemical traits determined by thirty-six genes were studied in various groups of drosophila. Continuous checking of phenetic data at all stages of the work by crossing various forms made this investigation not phenetic, as it usually happens in the case of population studies of a biochemical phenotype, but truly genetic. The study demonstrated the existence of at least five hierarchic levels within the group of species studied (populations, subspecies, "semispecies," sibling species, morphologically dissimilar species—see table 6.1)

The highest level of genetic and phenetic similarity (97%) was found in comparing the genetic characteristics of separate populations, and the lowest (35%), in comparing the same indices of morphologically different species within the whole group of closely related species.

This work demonstrated that subspecies are not a construct of systematicists, but a historically compounded complex of a population. They are more stable in both time and space than the separate populations composing them, and have a specific level of phenetic similarity (reflecting genetic similarity)—less than in two close populations, but notably greater than the similarity of even close species. From this it follows that in nature we may look for and identify traits of the subspecies rank. Phenetics makes it possible to do this by disclosing the *zones of stabilization and destabilization* of discrete traits.

In one of the phenogeographic works of Beregovoi and Danilov (1965), a study was made of the coloring of grouse (*Tetrao urogallus*). It developed that from the material at hand three different groups of grouse could be identified: two in the Urals and one in the European part of the USSR. In the Urals, from the south to the north gradual, insignificant changes were observed in intensity of the brown coloring (0.5 points per 6 degrees of latitude) to Sverdlovsk. Then a zone of very rapid change in the trait occurred (2 points per 3–4 degrees of latitude), and the zone of its stable state began (figure 6.10). As further analysis demonstrated, the grouse of the southern Urals on the one hand, and the Komi Republic and the northern Urals, on the other, differed also in dynamics of numbers and body size. The identified zones of stabilization included not one, but many populations

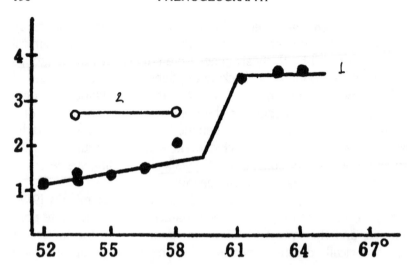

Figure 6.10. Cline variability in coloring of the grouse *Tetrao urogallus*, and the zone of "cline fracture" marking the boundary between subspecies: 1) Ural populations; 2) Belorussian, Smolensk, Leningrad, and Pskov populations. Vertical = coloring points; horizontal = geographic extent of area (Beregovoi and Danilov 1965).

of grouse. The rapid change in phenotype indicated the natural boundary between different population complexes—in this case, distinct subspecies.

Another example of the use of phenes for identifying subspecies pertains to the structural features of the second lower molar of a species of brush deermouse of North America. This species (*Peromyscus boylii*) lives in Mexico and the southern parts of the United States. Based on a whole series of traditional morphological traits (size and proportion of the body and the skull) and karyological traits (number of chromosome arms), three population complexes were identified. These complexes were clearly distinguishable according to the phenes of the structure of the second lower tooth. The American zoologist D. J. Schmidly, describing this case in 1973, noted that without considering the numerous other taxonomic traits, by this one tooth alone he could unmistakably identify *any* specimen from one of the population complexes, and with a high (but not 100 percent) degree of certainty identify animals from the other complexes. On the basis of the totality of all traits, the animals of the first complex were given species status, while the two other population complexes were identified as distinct subspecies within the other species.

Needless to say, not in all cases of studying intraspecific variability by phenogeographic methods can we expect to find distinct subspecies, historically compounded and differing from neighboring population complexes. This would be difficult to expect—first, because not all species have differentiated through evolution into subspecies, and second, because an investigator does not always notice the phenes that mark groups of populations close in evolution (figure 6.11) The task of the systematicist who uses phenogeographic methods is to find phenes in a concentration large enough to distinguish the given group from neighboring groups, reflecting a specific stage of phylogenesis, and not appearing independently in different parts of the species habitat.

One of the interesting problems of intraspecific taxonomy is that of "network relationship." At the intraspecific level much more often than at the species level, there are situations of polyphyletic origin of separate groups of the population even down to subspecies and semispecies. Phenes relating to the structure of the chromosome apparatus are particularly convenient for studying such situations: they can precisely indicate the paths and sequence of transformation of the genotype. More than forty variants of karyotype structure are known for the North American rodent, the pocket gopher (*Thomomys bottae*), twenty-nine subspecies of which inhabit California, Arizona, New Mexico, and the Mexican states of Sonora and Sinaloa (Patton 1972). According to the presence of a various number of pairs of different types of autosomes, both separate populations and whole population complexes, sometimes coinciding with subspecies, can be identified. It is interesting that based on the similarity of chromosome complements, it is possible to identify "primary" and "secondary" subspecies and to propose a basis for a hypothesis on the polyphyletic origin of some of these.

In intraspecific taxonomy many problems exist whose solution is linked to the development of phenetic methods of studying populations. How many levels of intraspecific hierarchy is it desirable

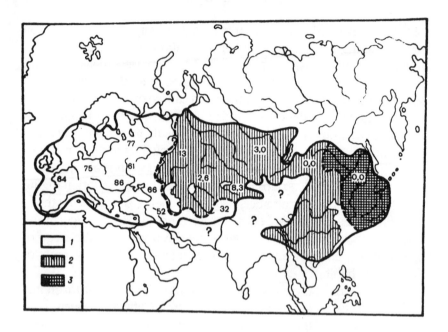

Figure 6.11. Phenogeography of the badger *Meles meles*: 1) distribution of the phene dark band on the head including the whole ear; 2) distribution of the phene narrow, dark band passing above the ear; 3) distribution of the phene dark brown coloring of the whole head. Numbers indicate percentages of incidence of the phene and presence of the first premolar tooth. (According to Geptner 1968, from Timofeeff-Ressovsky and Yablokov 1973)

to identify between the strictly population level and the species level? Is there a difference in the number of such levels in different groups of the animal and plant worlds? Many questions of this kind arise, and they can be answered only with the accumulation of data in the field of phenetics of natural populations.

Natural Selection
And Phenogeography

The study of natural selection is very important in the investigation of microevolution. Without a thorough understanding of the effect of this sole directional evolutionary factor, we cannot speak of any transition to managed evolution. Thus far, we know very few precise examples that disclose the effect of selection in nature; their number does not exceed several dozen.

Phenogeography gives the investigator useful equipment for studying selection. The example that has found its way into schoolbooks is the spread in England of the melanistic form of the birch measuring worm, Biston betularia—the phenomenon of industrial melanism disclosed and analyzed primarily on a phenogeographic basis. Another classic example of the effect of selection in which phenogeography played no small part was resolving why a high incidence of sickle-cell anemia in certain regions of Africa persisted, precisely in those regions where malaria was widespread. The gene for sickle-cell anemia is lethal in the homozygous state, and it would seem that it should become rare in a population. But in the heterozygous state, this gene establishes resistance to malaria, another horrible scourge in many regions of Africa. Selection maintains a certain frequency of the lethal sickle-cell gene and in this way decreases population losses due to another lethal illness. Both sickle-cell and resistance to malaria can be regarded as unique physiological phenes.

Some additional examples disclose the effect of selection through phenogeography. The study of the land snail, Cepaea vindobonensis, in the mountain valleys of Yugoslavia done at the end of the 1960s by the English geneticist J. Johnson disclosed an unusual distribution of phenes of banded shells and shells without bands. As it happened, the phene without bands occurs on slopes that are well warmed, while the snails with dark shells (which, as experiments proved, survive the cold better and use solar energy more efficiently) occupy the coldest slopes (figure 6.12).

Another species of the same genus, *Cepaea nemoralis*, was used in phenogeographic studies that disclosed the effect of selection. When the concentration of yellow shells was superimposed on a map of Europe, it turned out that there was a direct relation between the frequency of the yellow forms and the average summer temperature. It was established experimentally that the yellow morphs, as compared with the darker ones, have a lower survival rate under cold conditions.

One general conclusion can be drawn from all the thoroughly analyzed studies of the connection between geographic variability and the effect of selection: *wherever there is a cline variability of a trait, we can confidently assume that it is the effect of natural selection.*

The nutria (*Neofiber alleni*) that lives on the Florida peninsula exhibits a cline variability of pale yellow coloring and dark coloring. The percentage of pale yellow individuals in populations is very high in the northern region and very low in the south. The direct cause of this change in color is as yet unknown, but we may be certain that the main mechanism of the cline variability is the directional action of selection connected with some type of varying gradients of living conditions. An even sharper cline was described by M. Magomedmirzaev (1976) in his study of the change in proportion of red and brown seeds in the inflorescence of the willow-leafed rockrose (*Helianthemum*

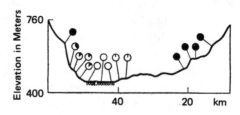

Figure 6.12. Distribution of land snail *Cepaea vindobonensis*, with and without bands, in a Yugoslavian valley. Dark color indicates incidence of individuals with dark shell living on the coldest slopes. (Jones, Leithe, and Rawlings 1977)

salicifolium) in a 250 km stretch along the western shore of the Caspian Sea. Ecological-physiological analysis showed a connection between the brown seed form and a higher xerophilousness—adaptation to drier living conditions—which, evidently, also determined the current pattern of distribution of these two phenes in this section of its habitat.

Two examples concern birds of the North Atlantic (Berry and Davis 1970). The great skua (*Stercorarius parasiticus*) lives in practically all suitable places on the shores of the North, Norwegian, Greenland, and Barents Seas. It has two coloring phenes, dark and pale yellow. The percentage of pale yellow birds is very small in the southern part of the habitat but reaches 100 percent in the north. Studies showed that the pale yellow form reaches sexual maturity earlier and is more aggressive. (This is important for successful feeding since the birds get a significant portion of their catch by taking it from small seagulls, which is probably why the species was given the Latin name *parasiticus*.) But in mating, the dark males have greater success. Finally, the dark males are more viable under conditions of a moderate climate, and the pale yellow, under Arctic Sea conditions. The complex combination of all these advantages also determines the appearance of a typical *balanced polymorphism*: not one of these forms has a decisive advantage over another over most of the habitat area (except for Greenland, where the dark form is completely absent).

Another example is the distribution of the "eye-glasses" phene in the thin-billed murre (*Uria aagle*). This phene is expressed in a dark circle around the eyes, joined over the bill by a thin, dark band. In the southernmost colony of the species, on the shores of Portugal, there are no eye-glass murres, but in the north, the number of such birds increases and reaches a maximum in the subarctic regions. Comparatively little is known about the directions of selection that maintain such a cline of variability (see Berry 1977). Disclosing these specific directions of selection is the work of physiologists, ecologists, and ethologists. Phenogeography in this case plays the role of a compass whose needle points to those situations in nature where the influence of selection can be clearly seen.

The last of this series of examples showing the potentials of phenogeography for disclosing the processes of natural selection in natural populations is the remarkable example of cline variability of the autumnal rustic moth (*Amathes glareosa*) on the Shetland Islands along the north shore of England (Kettlewell and Berry 1969). A study of this nocturnal moth was done by the English entomologist H. Kettlewell, who in the 1950s successfully studied industrial melanism in insects, specifically the well-known peppered moth. There are two forms of the autumnal rustic moth, the gray (A. *tipica*) and the dark, melanistic (A. *edda*). On the archipelago that stretches for 136 km, in the north there are only melanistic populations, and in the south, populations almost completely free of melanists. Changes in the frequency of melanists drawn on a graph showed a sharp break in the cline of variability in one of the regions of the main island where there is a small valley. Here, in 13 km, the concentration of the melanistic form changes by 35 percent (which corresponds on the average to 2.7 percent per kilometer), and on all the rest of the archipelago it changes on the average at 20 percent of this rate. Thus, phenogeography pointed to the most interesting place for further study. Such studies were organized and carried out in the vicinity of a valley with heather wastelands and occasional fields on limestone slopes.

    With this we could conclude the description of this case, since for further decoding of the natural situation, not phenogeography, but thorough ecological-genetic studies are needed. On both sides of the valley, 1–3 km wide in all, tagged moths were released and determinations were made as to where they would fly from the point of release. It developed that individual moths frequently fly more than 1½ km. A total of 1682 moths were tagged, and of these 65 were recovered. Regardless of the fact that constant winds blew during the experiment so that the moths had to be moved from one side of the valley to the other, only one moth was found that had flown to the opposite side of the valley.

    The experiment demonstrated that the valley is a real barrier to the mobility of the moths. Thus far we do not understand why this is so. Neither do we understand the adaptive significance of the darkening of the moths in the northern part of the archipelago. There are several hypotheses, specifically on the advantage of the camouflage coloring of the dark form under conditions of longer northern daylight. This is substantiated by the fact that in the stomachs of small gulls in the northern part of the archipelago, there were 20.7% light forms while the local natural population includes only 2.7% of the light forms. A difference was also detected in the lifespan under natural conditions between native individuals and those brought in from other places and released here. According to data from one experiment, the lifespan of local moths was 3.4 days, but those that were introduced, both dark and light forms, lived almost a day less. The well-known English saying "East or West, home is best" could not be proven better than in this case.

Studies of this moth showed that cline variability is favored by an unequal flow of migrants in the presence of a gradient of natural conditions, and *every cline represents a multitude of small, partially isolated populations of the little islands, with microdirections of natural selection that change,* depending on microconditions. Over large areas, the microdirections may be combined into notable macrophenogeographic changes.

On the whole, phenogeography offers broad possibilities for the study of natural selection, this key factor in the evolutionary process which in many ways is still a mystery. The main role that phenogeography plays here is the role of a "pointer" to situations that are promising for research whose purpose is to disclose the direction and pressure of selection. It is possible that in a number of cases, phenogeography will be of direct assistance in determining the direction of natural selection. This can be achieved by disclosing, for example, the correlation in distribution between some phene or group of phenes and some factors of the environment. These environmental factors may be abiotic (temperature, humidity, etc.) or they may be biotic, linked to other organisms. For example, a close tie was detected between the distribution of the phene of seven lateral large scales in the stickleback (*Gasterosteus aculeatus*) and the presence of predator fish in a reservoir. Certain phenes in the sand lizard correspond in their distribution to the habitat of the mountain ash in Euroasia. The problem of further ecological investigations is to elucidate the nature of this link, but its very existence is detected specifically by the phenogeographic approach.

Phenogeography as a Means
of Disclosing the Effect
of Evolutionary Factors

Among other evolutionary factors besides natural selection, which has been considered in detail above, in some cases phenogeography makes it possible to evaluate the effect of waves of population density in their clear, specific manifestation—the "founder effect."

In eastern England, between the shore of the New Bedford canal and the road that follows the canal for several kilometers, there is a slope 20–30 m wide covered by dense grass. This biotope is densely populated by the snail *Cepaea nemoralis*. Flooding in 1948 washed away all the old population of molluscs, but as early as three years later, all suitable living places were again inhabited by the snails. A study of the concentration of yellow shells, made selectively in 1952 on 200 m, showed a mixed pattern: various groups, even those that were close to each other, differed greatly in the frequency of this phene. The significance of the yellow coloring of the shell is clearly adaptive in certain habitats (camouflage coloring is effective in cases of visual selection by predators), but, evidently, coloring is quite neutral in the given conditions.

The mixed phenogeographic pattern of 1952 can be explained only by the founder effect in the settling of a free biotope. It is interesting that repeated studies carried out over nineteen years (three or four generations of molluscs) have not disclosed even the smallest difference in the frequency of phene distribution (Goodhart 1962, 1973, after Berry 1977). A similar case of the founder effect was also discovered when this species of mollusc settled in a completely new territory recently reclaimed in the polders near the city of Groningen.

It is much more difficult to identify the effect of isolation as an evolutionary factor by means of phenogeography. At first, this may seem very strange: what other devices, if not geographic, can we use to identify the action of such a primarily spatial factor as isolation? But it is not by chance that so little precise data exists about the effect of isolation on natural populations. First, it is very difficult to separate the effect of isolation from the effect of other factors of evolution, and second, the scale of the effect of isolation is usually not comparable to the scale of the lives of the investigators. Isolation may act uninterruptedly for hundreds of thousands of generations, sometimes weakening, sometimes increasing in strength.

The snail *Cepaea nemoralis* is a widely distributed species and one of the most thoroughly studied by evolutionists. It forms dense populations, has low mobility (average radius of individual activity does not exceed several dozen meters), and has a number of clear, genetically simply determined and easily identified traits, phenes. But how surprised scientists were to find this large and only slightly mobile snail on the cupola of St. Peter's Cathedral in Rome! It could have gotten there only on a strong gust of wind or have been brought by birds. In either case, the radius of the individual activity of this particular individual exceeded the average radius of activity of the species by hundreds of times. Genetic population experiments, like models of isolation in genetic processes, indicate that even a small change by genotypes (in general, even single percentage points) is capable of nullifying the genetic differences between initially different populations.

A third evolutionary factor, whose effect might lend itself to elucidation by phenogeographic methods, is the mutation process. The methodology in this case may be the following: If in the distribution of phenes, we find a concentration of rare, unusual phenes in any specific part of the habitat range, we may assume an intensive mutation process here. Of course, we are still far from precise proof; we may only assume that the proof will consist, as is usual in biology, in the conversion of an equation with many unknowns into equations with fewer and fewer unknowns.

So far, unfortunately, there is not a single well-analyzed example of the connection between the mutation process and obvious phenes, but in principle such a connection must exist, and finding it presents an engaging problem for future investigations. The main direction of such studies is locating centers of unusual phenes and subsequently identifying the possible effect of natural selection, waves of population density, isolation, and the results of combining identical alleles in a single genotype with an increased probability of pairing of related individuals (the effect of homozygotization).

Methods of Phenogeography

The principal method of phenogeography is to compare phene pools. This can be done with or without a map. In both cases the analysis is carried out by two methods—according to one phene or simultaneously according to a complex of phenes. Often a comparison based on one phene precedes further analysis of the complex of phenes, but sometimes it may be of microevolutionary interest in itself.

In the case of comparison centered around one phene, a map is made of the concentration of that phene at different points of the investigation, or only the presence of the given phene in a particular part of the habitat is noted. Depending on the problem of the study, one representation may be as convenient as the other.

Figure 6.13 shows the study by Timofeeff-Ressovsky of the distribution of the elaterii phene that governs the merging of spots on the first wing of the plant-eating ladybug. It is apparent that this phene occupies a large section within the habitat of the species and

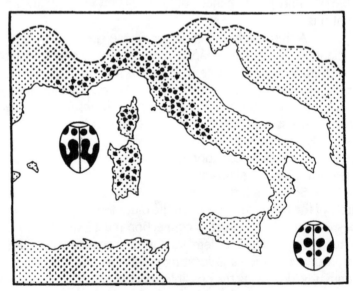

Figure 6.13. Distribution of the elaterii phene of merging of the spots on the first wings (large black spots) in the plant-eating ladybug, *Epilachna chrysomelina*, in the Mediterranean. (Timofeeff-Ressovsky 1940)

seems to have its own habitat. In this approach, the fact of the distribution of this phene within the limits of the habitat of the species, rather than its frequency, is interesting in itself. Such maps can disclose adaptive boundaries of phene distribution, find correlations in phene distrbution with other factors of the environment, and so on.

The study of distribution of the melanism phene in the hamster, *Cricetus cricetus*, in the Ukraine, carried out in the late 1930s by the well-known Soviet geneticist S. M. Gershenson (1945), is an example of such work. He was able to explain that the distribution of melanism is linked to the increased humidity of the habitat. But studies analyze phene distribution more often than noting the fact of simple presence or absence of a given phene in one section of a habitat or another. The case of the distribution of the simplex mutation (phene) in the tooth structure of the common vole in central Europe was mentioned earlier. From the zone of increased concentration of this phene in Jutland, there is, as it were, an outflow of traits with a gradually decreased concentration of this phene. Figure 6.14 shows the distribution of the phene of an interrupted band along the center of the back of the sand lizard within the range of its habitat. It is apparent that zones of increased concentration of the phene are quite sharply separated from zones of decreased concentration. There appear to be definite boundaries to its decreased or increased concentration. On analogous maps, constructed for the sand lizard corresponding to other phenes, very interesting boundaries appear. For some phenes, the Volga River, for example, is an impassable barrier; other phenes are found on its western bank in a high concentration and in a very low concentration on the eastern bank. For certain other phenes, however, no such barrier exists.

Earlier, it was emphasized repeatedly that not every analysis of a single phene will be meaningful and interesting if only because each phene has a certain scale in which its regular distribution will be evident. A visual example is the distribution of the phene of yellow shell coloring in the snail *Cepaea nemoralis* studied by the English geneticist J. Jones during 1974–1977. On the scale of a 5-km expanse, the distribution of yellow shells was random, and on the scale of an area of 22,500 km^2, no patterns were discerned. Also random was the distribution of the phene over all Great Britain. But on the scale of

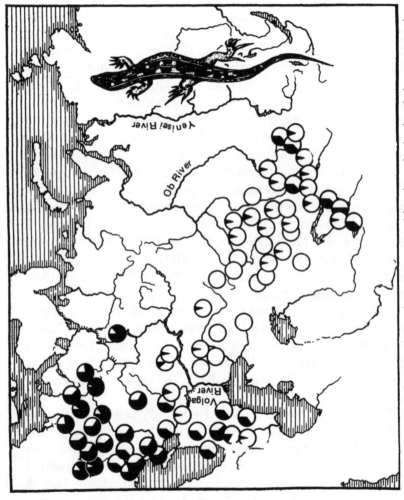

Figure 6.14. Distribution of the phene "interrupted band on the back" (dark sector) in the greater part of the habitat of the sand lizard *Lacerta agilis*. (Drawn by A. S. Baranov)

Europe, the yellow coloring phene seemed to have a cline distribution; that is, it was subject to selection and clearly adaptive.

Before considering the ways of comparing phene pools according to many phenes, I must digress into the area of mechanical psychology. I have said above that up to nine symbols is optimal for simultaneous perception by the human eye. This fact must be kept in mind when we map several phenes simultaneously. It is best if the map has no more than seven different symbols; otherwise it will be difficult to read.

Various methods of coding traits also diverge sharply in how effectively they are perceived visually. Individual phenes on a map may be designated by numbers, letters, or signs of varying configurations (triangle, square, rhombus, circle, etc.), and also by different colors. Experiments have shown that color is most effective for showing the location of a phene on a map.

When analyzing one or several phenes, we may place them on a map without considering the degree of concentration (see figure 6.3), but most widely used is a presentation of several phenes together with a consideration of their concentration. The concentration can be indicated by the size of the symbol, density of symbol distribution, or, if trait frequencies are being plotted, in a sum making up 100 percent, some variant of a sector graph. If we want to compare several independent traits from different groups of phenes, however, undoubtedly the best method is to use the wind rose (see figures 6.3 and 6.9).

As one of the methods in phenographic research methodology, we may consider computing a generalized phenetic index. Most widely used in current works, perhaps, is the comparison of separate populations and groups according to the so-called coefficient of coincidence ("measure of divergence," "index of similarity"; see chapter 5). The data obtained can be compared directly on a map or in tables, or it may serve as a basis for constructing a dendrogram portraying the phenetic similarity of the groups being studied. This method gives very good results and deserves to be widely used.

Another, as yet little used, method of comparison based on a generalized index is the simple method of index comparison. It can be successfully used to disclose intermediate forms between two or three already known, more extreme forms. A characteristic result of this method is the detection of hybrid zones between subspecies, semispecies, and species. The essence of the method may be illustrated by the exhaustive study of two semispecies of the American oriole carried out by the American zoologist J. D. Rising (figure 6.15).

On the Atlantic shore of North America, there are Baltimore orioles *Icterus galbula*, and on the Pacific shore, *I. bullockii*; in the central part of the continent from south to north, a broad hybrid zone exists between these semispecies. The traits of coloring were divided into nine separate groups (coloring of the head, neck, ear area, throat, etc.). Each group contained from three to five discrete color variants, phenes placed in order of increasing value of arbitrary points for each

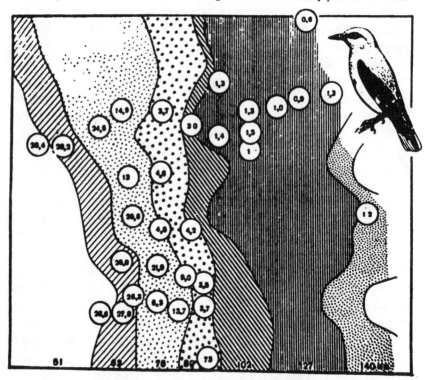

Figure 6.15. The correlation of averaged characteristics of coloring in indices (numbers in circles) with average yearly precipitation (numbers along horizontal axis) for two forms of the American oriole, *Icterus galbula* and *I. bullockii*, at the point of contact of their habitats in the central part of North America. For a typical *I. galbula*, the value of the index in points is 0; for a typical *I. bullockii*, 29. (After Rising 1970, from Timofeeff-Ressovsky and Yablokov 1973)

phene, from traits of the typical Baltimore oriole, taken as O, to traits characteristic for the Bullock oriole. The diversity of the summarized coloring indices for all traits exhibits a wide hybridization zone with a cline character of variability that correlates with the amount of precipitation. An analogous comparison may also be made in the form of a table without a map.

It is not accidental that the story of phenogeography was the longest in this book. Phenogeography is one of the most significant areas of phenetics since it makes it possible to pose and resolve the most complex problems of population research, from detecting population boundaries to reconstructing the course of evolution, the study of the microphylogenesis of a species. Phenogeography is important for the proper organization of fish and game management and a number of other biotechnical practices. Knowing the precise distribution of agriculturally important traits within the limits of a species habitat is a dependable basis for planning biotechnical procedures. Reestablishing the sable, *Martes zibellina*, in our country represents an excellent example of this type; this was achieved by catching and releasing sables from better stocks (populations and groups of populations characterized by an exceptionally successful complement of economically important traits—dark color, thickness of fur, large size).

Phenogeography is also very important in resolving the main problem of intraspecific systematics—the identification of groups of populations as separate subspecies.

Finally, phenogeography is of great significance in developing the theory of microevolution, which involves studying the mechanisms of divergence, spatial tendencies in the development of intraspecific variability, the formation of centers of diversity, cline variability, etc.

REFERENCES

Ayala, F. J., M. L. Tracey, D. Hedgecock, and R. C. Richmond. 1974. Genetic differentiation during the speciation process in *Drosophila*. *Evolution* 28:576–592.

Beregovoi, V. E. 1971. A study of polymorphism as a means of recognizing the chorological structure of species. (In Russian.) *Zh. Obshch. Biol.* 3(2):143–151.

Beregovoi, V. E. 1972. Analysis of polymorphism and quantitative evaluation of variability in populations, using the common spittlebug, *Philaenus spumarius* (L.), as an example. (In Russian.) *Zh. Obshch. Biol.* 33(6):

Beregovoi, V. E. and N. N. Danilov. 1965. Intraspecific variability of birds and pheno-geography. In *Intraspecific Variability of Land Animals and Microevolution*, pp. 166–174. (In Russian.) Sverdlovsk.

Berry, R. J. 1977. *Inheritance and Natural History*. London: Collins.

Berry, R. J. and P. E. Davis. 1970. Polymorphism and behavior in the Arctic skua (*Stercorarius parasiticus*). *Proc. Roy. Soc.*, ser. B, 175:255–267.

Bjorke, Tore. 1974. Geografisk sangvariasjon hos rodvingetrost, *Turdus iliacus*. *Sterna* 13(2):65–76.

Brussard, P. F. 1975. Geographic variation in North American colonies of *Cepaea nemoralis*. *Evolution* 29(3):402–410.

Carson, H. L. 1970. Chromosome traces of the origin of species. *Science* 158(3938):1414–1418.

Dessauer, H. C., W. Fox, and F. H. Pough. 1962. Starch-gel electrophoresis of trans-ferrins, esterases, and other plasma proteins of hybrids between two subspe-cies of whiptail lizard (genus *Cnemidophorus*). *Copeia* 4:767–774.

Dreux, P. 1978. Gene frequencies in the cat population of Paris. *J. hered.* 58:89–92.

Gershenson, S. M. 1945. Evolutionary studies on the distribution and dynamics of melanism in hamsters. *Genetics*, 30:207–251.

Goodhart, C. B. 1962. Variation in a colony of the snail *Cepaea nemoralis* (L.). *J. Anim. Ecol.* 32:207–237.

Goodhart, C. B. 1973. A 16-year survey of *Cepaea* on the Hundred-Foot Bank. *Malacolgia* 14:327–331.

Grimes. 1974. Dialects and geographical variation in the song of the splendid sunbird *Nectarinia coccinigaster*. *Ibis* 116(3):314–329.

Highton, R. 1962. Revision of the North American salamanders of the genus *Ple-thonodon*. *Bull. Fla. St. Mus.* (Biol. Sci.) 6:235–367.

Jones, J. S. 1973. Ecological genetics of a population of the snail *Cepaea nemoralis* at the northern limit of its range. *Heredity* 31(2):201–211.

Jones, J. S., B. H. Leithe, and P. Rawlings. 1977. Polymorphism in *Cepaea*: A problem with too many solutions *Ann. Rev. Ecol. Syst.* 8:109–143.

Jorgensen, J. B., L. Lederberg, C. Crebs, and H. Singer. 1965. Anthropological studies in the Hindu Kush and the Punjab. *Folk* 6(2):37–52.

Kettlewell, H. B. D. and R. J. Berry. 1969. Gene flow in a cline. *Heredity* 24:1–14.

Klun, J. A. et al. 1975. Insect sex pheromones: Intraspecific pheromonal variability of *Ostrinia nubilalis* in North America and Europe. *Environm. Entomol.* 4(6): 891–894.

Mayr, E. 1970. *Populations, Species and Evolution*. Cambridge: Harvard University Press.

Patton. 1972. Possible genetic consequences of meiosis in pocket gopher (*Thomomys bottae*) populations. *Experientia* 28(5):593–595.

Promptov, A. N. 1930. Geographic variability in the song of the chaffinch (*Fringilla coeleba* L.) in conjunction with general problems of seasonal bird migrations. (In Russian.) *Russkii Zool. Zh.* 10(3):17–40.

Rising, J. D. 1970. Morphological variation and evolution in some North American orioles. *Syst. Zool.* 19:315–351.

Rychkov, Y. G. and V. A. Sheremetyeva. 1977. The genetic process in the system of ancient human isolates in North Asia. In G. A. Harrison, ed., *Population Structure and Human Variation*, pp. 11, 47–108. London: Cambridge University Press.

Schmidly, D. J. 1973. Geographic variation and taxonomy of *Peromyscus boylii* from Mexico and the southern United States. *J. Mamm.* 54(1):111–130.

Selander, R. K. 1970. Behavior and genetic variation in natural populations. *Amer. Zool.* 10(1):53–56.

Serebrovskii, A. S. 1927. Genetic analysis of a population of domestic chickens of Dagestan uplands. (In Russian.) *Zh. Eksp. Biol.* 3(1–4):62–146.

Shubin, N. G. and M. L. Sedakova. 1982. Epigenetic polymorphism of the skull of the Siberian mole (*Talpa altaica*). In A. V. Yablokov, ed., *Population Phenetics*, pp. 269–275. (In Russian.) Moscow: Nauka.

Sturtevant, A. H. and T. Dobzhansky. 1936. Inversion of the third chromosome of a wild race of *Drosophila pseudoobscura* and their use in the study of the history of the species. *Proc. Nat. Acad. Sci. USA*, 22:448–456.

Timofeeff-Ressovsky, N. V. and A. V. Yablokov. 1973. Phenes, phenetics, and evolutionary biology. (In Russian.) *Priroda* (Moscow), 5:40–51.

Todd, N. B., G. E. Glass, and D. Creel. 1976. Cat population genetics in the U.S. Southwest and Mexico. *Carnivore Genetics Newsletter* 3(1):43–54.

Voipio, P. 1969. Variation of postfrontal pileus in *Lacerta vivipara*. *Ann. Zool. Fennici* 6(2):209–213.

Yudin, B. S. 1977. Individual and population variability of the dental system of the Siberian mole. (In Russian.) *Trudy Biologicheskogo Instituta* (Novosibirsk), 31: 178–199.

Zakharov, I. A. and S. O. Sergievskii. 1978. A study of changes in the composition of the *Adalia punctata* population in Leningrad and its environs. (In Russian.) *Genetika*, 14(2):281–284.

Zubov, A. A. 1973. *Ethnic Odontology.* (In Russian.) Moscow: Nauka.

Zweifel, R. G. 1962. Analysis of hybridization between two subspecies of the desert whiptail lizard *Cnemidophorus tigris, Copeia* 4:749–766.

FOR ADDITIONAL READING

Ayala, F. J. 1975. Genetic differentiation during the speciation process. *Evol. Biol.* 8:1–78.

Beregovoi, V. E. 1978. Genogeography and phenogeography of animal populations. In *Physiological and Populational Ecology of Animals* 5(7):34–41. (In Russian.) Saratov: Saratov University Press.

Borodin, P. M., M. N. Bochkarev, I. S. Smirnova, and G. P. Manchenko. 1978. Mutant allele frequencies in domestic cat populations of six Soviet cities. *J. Heredity* 69:169–174.

Cain, A. J. and J. D. Currey. 1963. Area effects in *Cepaea*. *Phil. Trans. Roy. Soc.*, ser. B, 246:1–81.

Eremina, I. V. 1976. Geography and evolution of phenes of the pattern of the chewing surface of the first lower and third upper molars in the bank vole. In *Physiological and Populational Ecology of Animals* 3(5):82–95. (In Russian.) Saratov: Saratov University Press.

Geptner, V. G. 1968. Some theoretical aspects of the problem of subspecies, subspecies traits, and the boundaries of subspecies habitats, using the geographical variability of two palearctic mammalian species as an example. (In Russian.) *Trudy Zoologicheskogo Muzeya Moskovskogo Universiteta* 10:3–36.

Harrison, G. A., J. S. Weiner, J. M. Tanner, and W. A. Barnicot. 1964. *Human Biology: An Introduction to Human Evolution, Variation and Growth.* New York: Oxford University Press.

Khokhutkin, I. M. 1971. Polymorphism and population boundaries of land molluscs of the genus *Bradibaena*. (In Russian.) *Ekologiya* 4:73–80.

Lamotte, M. 1951. Recherches sur la structure génétique des populations naturelles de *Cepaea nemoralis* L. *Biol. Bull. Suppl.* 35:1–239.

Magomedmirzaev, M. M. 1976. Problems of morphological measurement and computation from the position of plant phenetics. (In Russian.) *Zh. Obshch. Biol.* 37(3):331–343.

Novozhenov, Y. I. 1982. Geographical variability and population structure of species. In A. V. Yablokov, ed., *Population Phenetics*, pp. 78–90. (In Russian.) Moscow: Nauka.

Novozhenov, Y. I., V. E. Beregovoi, and I. M. Khokhutkin. 1973. Detecting population boundaries in polymorphic species according to frequency of incidence of forms. In N. N. Vorontsov, ed., *Problems of Evolution* 3:252–260. (In Russian.) Novosibirsk: Nauka.

Selander, R. K., W. G. Hunt, and S. Y. Yang. 1969. Protein polymorphism and genetic heterozygosity in two European subspecies of the house mouse. *Evolution* 23(3):379–390.

Serebrovskii, A. S. 1929. Problems and methods of genogeography. (In Russian.) *Proceedings of the All-Union Genetics Congress* 3:71–74.

Timofeeff-Ressovsky, N. V. 1940. Mutations and geographical variation. In J. Huxley, ed., *The New Systematics*, pp. 73–136. Oxford: Oxford University Press.

Timofeeff-Ressovsky, N. V., A. V. Yablokov, and N. V. Glotov. 1973. *Outline of Population Theory.* (In Russian.) Moscow: Nauka.

Vavilov, N. I. 1927. Geographical patterns in the distribution of genes in cultivated plants. (In Russian and English.) *Trudy Prikladnoi Botaniki, Genetiki, i Selektsii*, 17(3):411–428.

Wolda, H. 1963. Natural populations of the polymorphic land snail *Cepaea nemoralis* (L.). *Arch. neerl. Zool.* 15:381–471.

Yablokov, A. V. and V. Y. Etin. 1965. Analysis of intraspecific differences in mammalian body color (using the Greenland seal as an example). (In Russian.) *Zool. Zh.* 7:1094–1097.

Zimmerman, K. 1935. Zur Rassenanalyse der mitteleuropaischen Feldmause. *Arch. Naturgesch.* 4:258–276.

Conclusion

The reason for writing this book was the rapid development of a new direction in research on natural populations, phenetics. In its most general form, phenetics is an application of the ideas and methods of genetics to any nongenetic study of natural populations (from the position of the zoologist and botanist, the ecologist and morphologist, the physiologist and ethologist). On the one hand, phenetics makes it possible to pose and resolve population problems that, had they been put earlier, could not have been explained even with quite extensive material. On the other, phenetics poses tasks and problems which could not have been set out before because no methods existed for resolving them.

Phenetics can resolve the contradiction between the practical impossibility of broad genetic studies of species and the methodological need to obtain such data. With phenetics scientists can describe and decode, relatively quickly, hundreds and thousands of various evolutionary situations in nature. This elucidation of the paths of microevolution will most probably become the foundation of the theory of managed evolutionary processes.

Having put on "genetic eyeglasses" but not limiting themselves to the problems of classical population genetics, investigators in most areas of field biology, and possibly in a number of areas of experimental biology will have new ways of resolving scientific problems by resorting to phenetic examples and methods. All this will be useful in solving the cardinal problem of modern biology, managing the process of evolution; without this, as I have already said at the beginning of the book, it is impossible to imagine a fully valuable future existence for humanity in earth's biosphere.

The extent to which the phenetic approach proves productive in specific investigations depends not only on the phenetic methods used. Phenetics provides a key to the analysis of microevolutionary events taking place in nature, hidden from short-term observations that use the common, classic methods. What kind of a door this key will open depends on the investigator.

The broad interdisciplinary basis of phenetics combined with the desirability for mankind's rapid transition from uncontrolled interference with evolution to managed evolution compel scientists to write about phenetics both in proceedings of scientific conferences and highly specialized monographs and in publications addressed to a wide circle of readers interested in the study of living nature.

In conclusion, I would like to add that among biologists who studied populations using nonmetrical, discrete traits, some undoubtedly do not see the necessity of regarding phenetics as a new direction in biology. I can understand such a position. Will phenetics be a comparatively autonomous branch of population biology, or will it become one of its new methods, merging into other methods? It is difficult to say now. We have a Russian proverb "Let us wait, let us see."

If this book calls the attention of the reader to some problems of evolutionary or population biology, the task of the author will be half finished. But if it awakens in the reader an interest in studying the discrete, alternative traits in natural populations and evokes a desire in persons who are not geneticists to "put on genetic eyeglasses," and markedly extends the circle of material studied by geneticists, then the task of the author will be accomplished.

Bibliography

NOTE: Because of the character of this edition, this bibliography cannot be considered exhaustive. It includes certain works that give examples of the direction of phenetic research that were not mentioned in the text. An adequately complete bibliography on phenetics now includes, it seems, approximately 6000 to 7000 titles.

Abilkasimova, T. A. 1983. The use of traits for characterizing populations of *Nemachilus stoliczkai* (Cypriniformes, Cobitidae). (In Russian, English summary.) *Zool. Zh.*, 62(9):1371–1381.

Baranov, A. S. 1978. Identifying phenes according to color in reptiles (using the sand lizard as an example). In *Physiological and Populational Ecology of Animals* 5(7):68–72. (In Russian.) Saratov: Saratov University Press.

Berry, A. C. 1975. Factors affecting the incidence of nonmetrical skeletal variants. J. *Anat.* 120:519–535.

Berry, R. J. and A. G Searl. 1963. Epigenetic polymorphism of the rodent skeleton. *Proc. Zool. Soc.* (London) 140(4):557–615.

Berry, R. J., M. E. Jakobson, and J. Peters. 1978. The house mice of the Faroe Islands: A study in microdifferentiation. J. *Zool.* (London) 185:73–92.

Bol'shakov, V. N. and A. G. Vasil'ev. 1978. Epigenetic polymorphism in a population of bank and tundra voles with different degrees of spatial isolation. In *Physiological and Populational Ecology of Animals* 5(7):110–116. (In Russian.) Saratov: Saratov University Press.

Bol'shakov, V. N., I. A. Vasil'ev, and A. G. Maleeva. 1980. *Morphotypical Variability in Teeth of Field Mice.* (In Russian.) Moscow: Nauka.

Cesnys, G. and S. Pavilonis. 1982. On the terminology of nonmetric cranial traits (Discreta). *Homo* 33(2–3):125–230.

Evans, W. E. and A. V. Yablokov. 1983. *Variability in Cetacean Color Pattern: A New Approach to the Study of Mammalian Coloration.* (In Russian; English summary.) Moscow: Nauka.

Gill, A. E. 1977. Polymorphism in an island population of the California vole *Microtus californicus. Heredity* 38(1):1–11.

Gladkova, T. D. 1972. Skin patterns in higher and lower primates. In *Man: Evolution and Intraspecific Differentiation*. Proc. *Moscow Soc. Natur.* 43:84–100.

Grigor'eva, A. D., T. A. Grunt, and T. G. Sarycheva. 1978. Paleontology in phenetics. In *Physiological and Populational Ecology of Animals* 5(7):22–29. (In Russian.) Saratov: Saratov University Press.

Guseva, I. S. 1971. Some features of the manifestation of genes of a typological model of the papillary pattern of human fingers. (In Russian.) *Genetika* 7(10):103–115.

Hartman, S. E. 1980. Geographic variation analysis of *Dipodomys ordii* using nonmetric cranial traits. *J. Mamm.* 61(3):436–448.

Katakura, H. 1974. Variation analysis of elytral maculation in *Henosepilachna vigintioctomaculata* complex (Coleoptera, Coccinellidae). *Zoology*, ser. b, 19(2):445–455. (Journal of the Faculty of Science, Hokkaido University.)

Khit', G. L. 1983. *Dermatoglyphics of the Peoples of the* USSR. (In Russian; English summary.) Moscow: Nauka.

Kogan, Z. M. 1979. *Exterior and Interior Traits in Chickens*. (In Russian.) Novosibirsk: Nauka.

Kokhmanyuk, F. S. 1981. The Colorado beetle as a model of microevolution. (In Russian.) *Priroda* 12:86–88.

Kokhmanyuk, F. S. and V. E. Gaiduk. 1979. A study of the structure and dynamics of animal populations using methods of phenetics. (In *Conditions and Prospects for the Development of Morphology: Materials for an All-Union Conference*, pp. 365–366. (In Russian.) Moscow: Nauka.

Larina, N. I. 1978. General problems and methods in phenetic research. In *Physiological and Populational Ecology of Animals*, 5(7):12–22. (In Russian.) Saratov: Saratov University Press.

Larina, N. I. 1978. Using certain phenetic methods in studying intrapopulation groups in mice and field mice. In *Physiological and Population Ecology of Animals* 4(6):69–78. (In Russian.) Saratov: Saratov University Press.

Maleeva, A. G. and V. N. Popova. 1975. Late Pleistocene water vole from a formation of "mixed fauna" of the central and southern region east of the Urals. In *Fauna of the Urals and Northern Europe*, pp. 80–102. (In Russian.) Sverdlovsk.

Mamaev, S. A. and A. K. Makhnev. 1982. A study of the population structure of wood plants by means of morphophysiological markers. In A. V. Yablokov, ed., KPhenetics of Populations, pp. 140–149. (In Russian.) Moscow: Nauka.

Marga Stewart, M. M. 1974. Parallel pattern polymorphism in the genus *Phrynobatrachus* (Amphibia: Ranidae). *Copeia* 4:823–832.

Nesterov, G. A. 1981. Individual recognition of fur seals by their dermatoglyphics. (In Russian.) *Priroda* 7:91–93.

Nevo, E. 1973. Adaptive color polymorphism in cricket frogs. *Evolution* 27(3):353–367.

Novozhenov, Y. I. 1978. Phenogeography of stable polymorphism. In *Physiological and Populational Ecology of Animals* 5(7):41–47. (In Russian.) Saratov: Saratov University Press.

Orlov, L. M. 1975. Venation of the wing of *Chrysopa aspersa* Wesm. (Chrysopidae, Neuroptera) as a model of microevolutionary research. (In Russian; English summary.) *Zh. Obshch. Biol.* 36(6):903–913.

Rees, J. W. 1980. Morphological variation in the cranium and mandible of the white-tailed deer (*Odocoileus virginianus*): A comparative study of geographical and four biological distances. *J. Morph.* 128:95–112.

Rychkov, Y. G. and A. A. Movsesyan. 1972. Genetic-anthropological analysis of the distribution of anomalies of the skull of mongoloids in Siberia in conjunction with the problem of their origin. In *Man: Evolution and Intraspecific Differentiation*. (In Russian.) *Proc. Moscow Soc. Natur.* 43:113–132.

Sarycheva, T. G. and A. V. Yablokov. 1973. Paleontology and microevolution. *Zh. Obshch. Biol.* 31(3):348–359.

Scott, P. *The Wild Swans of Slimbridge*, pp. 1–14. Wildfowl Trust, Slimbridge.

Soule, M. 1972. Phenetics of natural populations. III: Variations in insular populations of lizard. *Amer. Natur.* 106(950):429–446.

Svala, E. and O. Halkka. 1974. Geographical variability of front-oclypeal color polymorphism in *Philaenus spumaris* (L.) (Homoptera). *Ann. Zool. Fenn.* 11(4):283–287.

Thompson, V. and O. Halkka. 1973. Color polymorphism in some North American *Philaenus spumarius* (Homoptera:Aphrophoridae) populations. *Amer. Midl. Natur.* 89(2):348–359.

Todd, N. B., G. E. Glass, and I. McLure. 1974. Gene frequencies in some cats of South America. *Carnivore Genet. Newsl.* 2:230–235.

Tordoff, W. III, and D. Pettus. 1977. Temporal stability of phenotypic frequencies in *Pseudacris triseriata* (Amphibia, Anura, Hylidae). *J. Herpetol.* 11(2):161–168.

Turutina, L. V. and V. I. Podmarev. 1978. Territorial distribution and identification of genetic-spatial groups in a population of agile lizards. In *Physiological and Populational Ecology of Animals* 5(7):72–77. (In Russian.) Saratov: Saratov University Press.

Wiltafsky, H. 1973. Die geographische Variation morphologischer Merkmale bei *Sciurus vulgaris* L., 1758. Inaugural-Dissertation der Mathemat.-Naturwiss. Fakultät der Universität zu Köln, Köln.

Yablokov, A. V. 1978. A study of the spatial structure of a population of animals using analysis of frequency of incidence of discrete traits. In *Abstracts of Papers of the 14th International Genetics Congress*. Part 1, section 13–20:494.

Yablokov, A. V. 1982. *Phenetics of Populations*. (In Russian.) Moscow: Nauka.

Yablokov, A. V., W. F. Perrin, and M. V. Mina. 1983. Evaluation of phenetic relations among groups of dolphins, using analysis of nonmetric cranial variation. (In Russian, English summary.) *Zool. Zh.* 62(12):1887–1896.

Yakovlev, V. N., Y. G. Izumov, and A. N. Kas'yanov. 1981. The phenetic method in studying fish populations. (In Russian.) *Biol. Nauki* 2:98–101.

Zakharov, V. M. 1978. Basic methods of population studies of bilateral structures of animals. In *Physiological and Populational Ecology of Animals* 5(7):54–60. (In Russian.) Saratov: Saratov University Press.

Zubov, A. A. and N. I. Khaltseeva, eds. 1979. *Ethnic Odontology of the* USSR. (In Russian.) Moscow: Nauka.

Index

Printed in the USA
CPSIA information can be obtained
at www.ICGtesting.com
JSHW021321221024
72173JS00011B/1627